Corrosion Chemistry

Corrosion Chemistry

Volkan Cicek and Bayan Al-Numan

Scrivener

WILEY

Library of Congress Cataloging-in-Publication Data:

ISBN 978-0-470-94307-6

Printed in the United States of America

10 9 8 7 6 5 4 3 2 1

Contents

List of Tables

Acknowledgements

First of all, I would like to express my sincere appreciation and gratefulness to Wiley-Scrivener Publications since they were eager to fund such a project at a time, when almost all academic publications in Iraq ceased. We all hope that the negative situation in Iraq is temporary and Iraq will take its place in the academic world in the near future once more.

I also would like to acknowledge Dr. Bayan Al Numan for contributing to this book with a vast chapter "Corrosion in Engineering Materials," and graphic artist Khabbab Habib for his support when needed.

I also would like to acknowledge my Ph.D. advisor, Dr. Allen W. Apblett, for his guidance, motivation, and inspiration throughout my Ph.D. education. His valuable advice, criticism, and encouragement have greatly helped me. I have benefited much from his broad range of knowledge, and his scientific approach.

Thanks are also due to my father, my mom, my sister, my relatives, and friends for their moral support, and encouragement throughout the years.

Finally, I am deeply indebted to my wife, Mine Cicek, for her unconditional love, patience, care, and sacrifice. Your moral support during this time, together with the enthusiasm that I got from our two lovely children, Furkan and Zehra, was invaluable to me.

Preface

This book was written and published as a reference for engineers and a textbook for students, due to the necessity of such a book in the area of corrosion chemistry and corrosion science in general.

Corrosion is, in essence, a chemical process, and it is crucial to understand the dynamics from a chemical perspective before proceeding with analyses, designs and solutions from engineering aspect. The opposite is also true in the sense that scientists should take into consideration the contemporary aspects of the issue as it relates to the daily life before proceeding with specifically designed theoretical solutions. Thus, this book is advised to both theoreticians and practitioners of corrosion alike.

Corrosion costs billions of dollars to each and every single economy in the world however is only taught in the form of a single semester elective course and that is sometimes at undergraduate level since most probably it is a joint discipline that is associated with many others, thus does not belong to any major science altogether. Corrosion is associated primarily with major engineering sciences such as chemical engineering, civil engineering, petroleum engineering, mechanical engineering, metallurgical engineering, mining engineering among others and major fundamental sciences such as subdisciplines of physical, inorganic and analytical chemistry as well as physics and biology, e.g., electrochemistry, surface chemistry and physics, solution chemistry, crystalline and amorphous structures and solid state chemistry and physics in general, microbiology, etc.

Hence, a reference book that summarizes the process with its contemporary aspects with respect to both scientific and engineering aspects was needed. Additionally, being such a joint discipline such a book should not overwhelm the reader with too much detail but only enough to understand the process as this was aimed in this book.

In addition to be used as a reference, this book could be used as a textbook most conveniently for a single semester elective course; however, the period of the course could be adjusted to fit into a long or a short summer term as well as a complete year depending on the course. In the case that this book is used as a textbook for a full year course, using supplementary resources may be beneficial especially in the case of engineering sciences.

1

Corrosion and Its Definition

According to American Society for Testing and Materials' corrosion glossary, corrosion is defined as "the chemical or electrochemical reaction between a material, usually a metal, and its environment that produces a deterioration of the material and its properties".[1]

Other definitions include Fontana's description that corrosion is the extractive metallurgy in reverse,[2] which is expected since metals thermodynamically are less stable in their elemental forms than in their compound forms as ores. Fontana states that it is not possible to reverse fundamental laws of thermodynamics to avoid corrosion process; however, he also states that much can be done to reduce its rate to acceptable levels as long as it is done in an environmentally safe and cost-effective manner.

In today's world, a stronger demand for corrosion knowledge arises due to several reasons. Among them, the application of new materials requires extensive information concerning

1

corrosion behavior of these particular materials. Also the corrosivity of water and atmosphere have increased due to pollution and acidification caused by industrial production. The trend in technology to produce stronger materials with decreasing size makes it relatively more expensive to add a corrosion allowance to thickness. Particularly in applications where accurate dimensions are required, widespread use of welding due to developing construction sector has increased the number of corrosion problems.[3] Developments in other sectors such as offshore oil and gas extraction, nuclear power production and medicinal health have also required stricter rules and control. More specifically, reduced allowance of chromate-based corrosion inhibitors due to their toxicity constitutes one of the major motivations to replace chromate inhibitors with environmentally benign and efficient ones.

2

The Corrosion Process and Affecting Factors

There are four basic requirements for corrosion to occur. Among them is the anode, where dissolution of metal occurs, generating metal ions and electrons. These electrons generated at the anode travel to the cathode via an electronic path through the metal, and eventually they are used up at the cathode for the reduction of positively charged ions. These positively charged ions move from the anode to the cathode by an ionic current path. Thus, the current flows from the anode to the cathode by an ionic current path and from the cathode to the anode by an electronic path, thereby completing the associated electrical circuit. Anode and cathode reactions occur simultaneously and at the same rate for this electrical circuit to function.[4] The rate of anode and cathode reactions (that is the corrosion rate), is defined by American Society for Testing and Materials as material loss per area unit and time unit.[1]

In addition to the four essentials for corrosion to occur, there are secondary factors affecting the outcome of the corrosion reaction. Among them there are temperature, pH, associated fluid dynamics, concentrations of dissolved oxygen and dissolved salt. Based on pH of the media, for instance, several different cathodic reactions are possible. The most common ones are:

Hydrogen evolution in acid solutions,

$$2H^+ 2e^- \longrightarrow H_2 \qquad (2.1)$$

Oxygen reduction in acid solutions,

$$O_2 + 4H^+ 4e^- \longrightarrow 2H_2O \qquad (2.2)$$

Hydrogen evolution in neutral or basic solutions,

$$2H_2O + 2e^- \longrightarrow H_2 + 2OH^- \qquad (2.3)$$

Oxygen reduction in neutral or basic solutions,

$$O_2 + 2H_2O + 4e^- \longrightarrow 4OH^- \qquad (2.4)$$

The metal oxidation is also a complex process and includes hydration of resulted metal cations among other subsequent reactions.

$$M^0 \longrightarrow M^{n+} + ne^-, \qquad (2.5)$$

In terms of pH conditions, this book has emphasized near neutral conditions as the media leading to less emphasis on hydrogen evolution and oxygen reduction reactions, since both hydrogen evolution and oxygen reduction reactions that take place in acidic conditions are less common.

Among cathode reactions in neutral or basic solutions, oxygen reduction is the primary cathodic reaction due to the difference in electrode potentials. Thus, oxygen supply to the system, in which corrosion takes place, is of utmost importance for the outcome of corrosion reaction. Inhibitors are commonly tested in stagnant solutions for the purpose of weight-loss tests, thus ruling out the effects of varying fluid dynamics on corrosion. Weight-loss tests are performed at ambient conditions, thus effects of temperature and dissolved oxygen amounts are not tested as well, while for salt fog chamber tests, temperature is varied for accelerated corrosion testing. For both weight loss tests and salt fog chamber tests, however, dissolved salt concentrations are commonly kept high for accelerated testing to be possible.

When corrosion products such as hydroxides are deposited on a metal surface, a reduction in oxygen supply occurs, since the oxygen has to diffuse through deposits. Since the rate of metal dissolution is equal to the rate of oxygen reduction, a limited supply and limited reduction rate of oxygen will also reduce the corrosion rate. In this case the corrosion is said to be under cathodic control.[5] In other cases corrosion products form a dense and continuous surface film of oxide closely related to the crystalline structure of metal. Films of this type prevent primarily the conduction of metal ions from metal-oxide interface to the oxide-liquid interface, resulting in a corrosion reaction that is under anodic control.[5] When this happens, passivation occurs and metal is referred as a passivated metal. Passivation is typical for stainless steels and aluminum.

3

Corrosion Types Based on Mechanism

Brief definitions of major types of corrosion will be given in this section in the order of commonalities and importance of these corrosion types for the metal alloys, which are mild steel, and Aluminum 2024, 6061 and 7075 alloys.

3.1 Uniform Corrosion

Uniform corrosion occurs when corrosion is quite evenly distributed over the surface, leading to a relatively uniform thickness reduction.[6-7] Metals without significant passivation tendencies in the actual environment, such as iron, are liable to this form. Uniform corrosion is assumed to be the most common form of corrosion and responsible for most of the material loss.[6] However, it is not a dangerous form of corrosion because prediction of thickness reduction rate can be done by means of simple tests.[7] Therefore, corresponding corrosion allowance can be added, taking into account strength requirements and lifetime.

3.2   Pitting Corrosion

Pitting corrosion is one of the most observed corrosion types for aluminum and steel, and it is the most troublesome one in near neutral pH conditions with corrosive anions, such as Cl⁻ or $SO4^{2-}$ present in the media.[8–11] It is characterized by narrow pits with a radius of equal or lesser magnitude than the depth. Pitting is initiated by adsorption of aggressive anions, such as halides and sulfates, which penetrate through the passive film at irregularities in the oxide structure to the metal-oxide inter-face. It is not clear why the breakdown event occurs locally.[9] In the highly disordered structure of a metal surface, aggres-sive anions enhance dissolution of the passivating oxide. Also, adsorption of halide ions causes a strong increase of ion con-ductivity in the oxide film so that the metal ions from the metal surface can migrate through the film.

Thus, locally high concentrations of aggressive anions along with low solution pH values strongly favor the process of pit-ting initiation. In time, local thinning of the passive layer leads to its complete breakdown, which results in the formation of a pit. Pits can grow from a few nanometers to the microm-eter range. In the propagation stage, metal cations from the dissolution reaction diffuse toward the mouth of the pit or crevice (in the case of crevice corrosion), where they react with OH⁻ ions produced by the cathodic reaction, forming metal hydroxide deposits that may cover the pit to a varying extent. Corrosion products covering the pits facilitate faster corrosion because they prevent exchange of the interior and the exterior electrolytes, leading to very acidic and aggressive conditions in the pit.[9–11] Stainless steels have high resistance to initiation of pitting. Therefore, rather few pits are formed, but when a pit has been formed, it may grow very fast due to large cathodic areas and a thin oxide film that has considerable electrical conductance.[12] Conversely for several aluminum alloys, pit initiation can be accepted under many circumstances. This is so because numerous pits are formed, and the oxide is insulating and has, therefore, low cathodic activity. Thus, corrosion rate is under cathodic control. However, if the cathodic reaction can occur on a different metal because of galvanic connection as for deposition of Cu on the aluminum surface, pitting rate may be very high. Therefore, the nature of alloying elements is very important.[13]

3.3   Crevice Corrosion

Crevice corrosion occurs underneath deposits and in narrow crevices that obstruct oxygen supply.[14–16] This oxygen is initially required for the formation of the passive film and later for repassivation and repair. Crevice corrosion is a localized corrosion concentrated in crevices in which the gap is wide enough for liquid to penetrate into the crevice but too narrow for the liquid to flow. A special form of crevice corrosion that occurs on steel and aluminum beneath a protecting film of metal or phosphate, such as in cans exposed to atmosphere, is called filiform corrosion.[14] Provided that crevice is sufficiently narrow and deep, oxygen is more slowly transported into the crevice than it is consumed inside it. When oxygen has been completely consumed, OH$^-$ can no longer be produced there. Conversely, dissolution of the metal inside the crevice continues, driven by the oxygen reduction outside of the crevice. Thus, the concentration of metal ions increases and, with missing OH$^-$ production in the crevice, electrical neutrality is maintained by migration of negative ions, such as Cl$^-$, into the crevice.[15] This way, an increasing amount of metal chlorides or other metal salts are produced in the crevice. Metal salts react with water and form metal hydroxides, which are deposited, and acids such as hydrochloric acid, which cause a gradual reduction of pH down to values between 0 and 4 in the crevice, while outside of crevice it is 9 to 10, where oxygen reduction takes place. This autocatalytic process leads to a critical corrosion state. Thus reduction of hydronium ions takes place in very acidic conditions in addition to the primary cathodic reaction that is reduction of oxygen[16]

$$2H^+ + 2e^- \longrightarrow H_2 \qquad (3.1)$$

$$O_2 + 2H_2O + 4e^- \longrightarrow 4OH^- \qquad (3.2)$$

3.4   Galvanic Corrosion

Galvanic corrosion occurs when a metallic contact is made between a more noble metal and a less noble one.[17–19] A necessary condition is that there is also an electrolytic condition between the metals, so that a closed circuit is established. The area ratio between cathode and anode is very important. For instance, if the more noble cathodic metal has a large surface area and the less noble metal has a relatively small area, a large cathodic reaction must be balanced by a correspondingly large anodic reaction concentrated in a small area, resulting in a higher anodic reaction rate.[17] This leads to a higher metal dissolution rate or corrosion rate. Therefore, the ratio of cathodic to anodic area should be kept as low as possible. Galvanic corrosion is one of the major practical corrosion problems of aluminum and aluminum alloys,[18] since aluminum is thermodynamically more active than most of the other common structural materials and the passive oxide, which protects aluminum, may easily be broken down locally when the potential is raised due to contact with a more noble material. This is particularly the case when aluminum and its alloys are exposed in waters containing chlorides or other aggressive species.[19]

The series of standard reduction potentials of various metals can be used to explain the risk of galvanic corrosion; however, these potentials express thermodynamic properties, which do not take into account the kinetic aspects.[20] Also, if the potential difference between two metals in a galvanic couple is too large, the more noble metal does not take part in the corrosion process with its own ions. Thus, under this condition, the reduction potential of the more noble metal does not play any role. Therefore, establishing a galvanic series for specific conditions becomes crucial. For example, a new galvanic series of different materials in seawater at 10°C and at 40°C has been established by University of Delaware Sea Grant Advisory Grant Program,[18] and a more detailed one by the Army Missile Command.[21] According to these galvanic series, Aluminum 6061-T6 alloy is more active than 7075-T6 alloy, which is more active than 2024-T4 alloy. In this scheme, mild steel ranks lower than the aluminum alloys. This order may be opposite to the order of corrosion affinity in different circumstances, such as in the case for aircrafts.[21]

3.5   Intergranular Corrosion

Intergranular corrosion is the localized attack with propagation into the material structure with no major corrosion on other parts of the surface.[6,22–25] The main cause of this type of corrosion is the presence of galvanic elements due to differences in concentration of impurities or alloying elements.[6] In most cases, there is a zone of less noble metal at or in the grain boundaries, which acts as an anode, while other parts of the surface form the cathode.[22] The area ratio between the cathode and anode is very large and, therefore, the corrosion rate can be high. The most familiar example of intergranular corrosion is associated with austenitic steels.[23] A special form of intergranular corrosion in aluminum alloys is exfoliation corrosion.[24] It is most common in AlCuMg alloys, but it is also observed in other aluminum alloys with no copper present. Both exfoliation corrosion and other types of intergranular corrosion are efficiently prevented with a coating of a more resistant aluminum alloy, such as an alclad alloy or commercially pure aluminum, which is the reason alclad 2024-T3 alloy is used in most modern aircrafts.[25]

3.6   Selective Corrosion

Selective corrosion or selective leaching occurs in alloys in which one element is clearly less noble than the others.[26] As a result of this form of corrosion; the less noble metal is removed from the material, leading to a porous material with very low strength and ductility. However, regions that are selectively corroded are sometimes covered with corrosion products or other deposits. Thus, the component keeps exactly the same shape, making the corrosion difficult to discover.[26]

3.7   Erosion or Abrasion Corrosion

Erosion or abrasion corrosion occurs when there is a relative movement between a corrosive fluid and a metallic material immersed in it.[6,27] In such cases, the material surface is exposed to mechanical wear, leading to metallically clean surfaces, which results in a more active metal. Most sensitive materials are those normally protected by passive oxide layers with inferior strength and adhesion to the substrate, such as lead, copper, steel and some aluminum alloys. When wearing particles move parallel to the material surface, the corrosion is called abrasion corrosion. On the other hand, erosion corrosion occurs when the wearing particles move with an angle to the substrate surface.[27]

3.8   Cavitation Corrosion

Cavitation corrosion occurs at fluid dynamic conditions, causing large pressure variations due to high velocities, as often is the case for water turbines, propellers, pump rotors and external surfaces of wet cylinder linings in diesel engines.[6,22–23] While erosion corrosion has a pattern reflecting flow direction, cavitation attacks are deep pits grown perpendicularly to the surface. Pits are often localized close to each other or grown together over smaller or larger areas, making a rough, spongy surface.[23]

3.9   Fretting Corrosion

Fretting corrosion occurs at the interface between two closely fitting components when they are subjected to repeated slight relative motion.[23,28] The relative motion may vary from less than a nanometer to several micrometers in amplitude. Vulnerable objects are fits, bolted joints and other assemblies where the interface is under load.[28]

3.10   Stress Corrosion Cracking

Stress Corrosion Cracking is defined as crack formation due simultaneous effects of static tensile strength and corrosion.[23,29] Tensile stress may originate from an external load, centrifugal forces, temperature changes or internal stress induced by cold working, welding or heat treatment. The cracks are generally formed in planes normal to the tensile stress, and they propagate intergranularly or transgranularly and may be branched.[29]

Corrosion fatigue is crack formation due to varying stresses combined with corrosion.[23,30] This is different from stress corrosion cracking because stress corrosion cracking develops under static stress while corrosion fatigue develops under varying stresses.[30]

3.11   Microbial Corrosion

Another type of corrosion occurs when organisms produce an electron flow, resulting in modification of the local environment to a corrosive one.

An example is when microbial deposits accumulate on the surface of a metal. They can be regarded as inert deposits on the surface, shielding the area below from the corrosive electrolyte. The area directly under the colony will become the anode, and the metallic surface just outside the contact area will support the reduction of oxygen reaction and become the cathode. Metal dissolution will occur under the microbial deposit and, in that regard, would resemble to pits, but the density of local dissolution areas should match closely with the colony density.

Another case is when microbial deposits produce components, such as inorganic and organic acids, that will change the local environment and thereby induce corrosion. Specifically, the production of inorganic acids leads to hydrogen ion production, which may contribute to hydrogen embrittlement of the colonized metal.

In anaerobic conditions, some bacteria can reduce the sulfate ion to produce oxygen and the sulfide ion. The sulfide ion then combines with ferrous ions to form iron sulfide, leading to the dissolution of the metal surface. Some other bacteria can directly reduce metal atoms to ions. Impedance spectroscopy is one test technique that is applicable to biocorrosion. Potentiodynamic scans may be used to determine the effect of biofilms in both anodic and cathodic behavior.[468]

4

Corrosion Types of Based on the Media

Corrosion types can also be categorized based on what type of environment they take place. Accordingly, major corrosion types are atmospheric corrosion, corrosion in fresh water, corrosion in seawater, corrosion in soils, corrosion in concrete and corrosion in the petroleum industry.

4.1   Atmospheric Corrosion

In general for atmospheric corrosion, dusts and solid precipitates are hygroscopic and attract moisture from air. Salts can cause high conductivity, and carbon particles can lead to a large number of small galvanic elements since they act as efficient cathodes after deposition on the surface.[32,33] The most significant pollutant is SO_2, which forms H_2SO_4 with water.[34,35] Water that is present as humidity bonds in molecular form to even the cleanest and well-characterized metal surfaces.[32,33]

Through the oxygen atom it bonds to the metal surface or to metal clusters and acts as a Lewis base by adsorbing on electron deficient adsorption sites. Water may also bond in dissociated form, in which case the driving force is the formation of metal-oxygen or metal-hydroxyl bonds. The end products resulting from water adsorption are then hydroxyl and atomic hydrogen groups adsorbed on the substrate surface.[36] Atmospheric corrosion rate is influenced by the formation and protective ability of the corrosion products formed. The composition of corrosion products depends on participating dissolved metal ions and anions dissolved in the aqueous layer. According to the hard and soft acids and bases theory, hard metal ions such as Al^{3+} and Fe^{3+} prefer H_2O, OH^-, O^{-2}, SO_4^{-2}, NO_3^-, CO_3^{-2} while intermediate metals such as Fe^{2+}, Zn^{2+}, Ni^{2+}, Cu^{2+}, Pb^{2+} prefer softer bases, such as SO_3^{-2} or NO_2^- and soft metals such as Cu^+ or Ag^+ prefer soft bases as R_2S, RSH or RS^-.[34–35]

In the specific case of iron or steel exposed to dry or humid air, a very thin oxide film composed of an inner layer of magnetite (Fe_3O_4) forms, covered by an outer layer of FeOOH (rust).[37–38] Atmospheric corrosion rates for iron are relatively high and exceed those of other structural metals. They range (in μm/ year) from 4 to 65 in rural, 26 to 104 in marine, 23 to 71 in urban and 26 to 175 in industrial areas.[39]

In the case of aluminum, the metal initially forms a few nm thick layer of aluminum oxide, γ-Al_2O_3, which in humidified air is covered by aluminum oxyhydroxide, γ-AlOOH, eventually resulting in a double-layer structure.[40–42] The probable composition of the outer layer is a mixture of Al_2O_3 and hydrated Al_2O_3, mostly in the form of $Al(OH)_3$. However, the inner layer is mostly composed of Al_2O_3 and small amounts of hydrated aluminum oxide mostly in the form of AlOOH.[43–45] This oxide layer is insoluble in the pH interval of 4 to 9.[46] Lower pH values results in the dissolution of Al^{3+}. Rates of atmospheric corrosion of aluminum outdoors (in μm/year) are substantially lower than for most other structural metals and are from 0.0 to 0.1 in rural, from 0.4 to 0.6 in marine, and ~1 in urban areas.[47, 48]

In general, anodic passivity of metals, regardless of type of corrosion, is associated with the formation of a thin oxide film, which isolates the metal surface from the corrosive environment. Films with semiconducting properties, such as Fe, Ni, Cu oxides, provide inferior protection compared to metals as Al, which has an insulating oxide layer.[49]

An alternative explanation of differences between oxide films of different metals based on their conducting properties is the networkforming oxide theory, in which covalent bonds connect the atoms in a three-dimensional structure. Due to nature of covalent bonding, there is short-range order on the atomic scale, but no long-range order. These networks of oxides can be broken up by the introduction of a network modifier.[50] When a network modifier is added to a networkforming oxide, they break the covalent bonds in the network, introducing ionic bonds, which can change the properties of mixed oxides, such as Cu/Cu_2O or Al/Al_2O_3, where rate of diffusion of Cu in Cu_2O is 10,000 times larger than Al in Al_2O_3.[51] Depending on single oxide bond strengths, metal oxides can be classified as network formers, intermediates or modifiers. Network formers tend to have single oxide strengths greater than 75 kcal/mol, intermediates lie between 75 and 50 and modifiers lie below this value.[52,53] Iron is covered by a thin film of cubic oxide of γ-Fe_2O_3/Fe_3O_4 in the passive region. The consensus is that the γ-Fe_2O_3 layer, as a network former, is responsible for passivity, while Fe_3O_4, as a network modifier, provides the basis for formation of higher oxidation states but does not directly contribute toward passivity.[54] The most probable reason for iron being more difficult to passivate is that it is not possible to go directly to the passivating species of γFe_2O_3. Instead, a lower oxidation state film of Fe_3O_4 is required, and this film is highly susceptible to chemical dissolution. Until the conditions are established whereby the Fe_3O_4 phase can exist on the surface for a reasonable period of time, the γ-Fe_2O_3 layer will not form and iron dissolution will continue.[55–56] Impurities such as water also modify the structure of oxide films. Water acts as a modifying oxide when added to network-forming oxides and thus weakens the structure.[57,58] In conclusion, metals, which fall into network-forming or intermediate classes, tend to grow protective oxides, such as Al or Zn. Network formers are non-crystalline, while the intermediates tend to be micro-crystalline at low temperatures. The metals, which are in the modifier class, have been observed to grow crystalline oxides, which are thicker and less protective.[59] A partial solution is to alloy the metal with one that forms a network-forming oxide, in which the alloying metal tends to oxidize preferentially and segregates to the surface as a glassy oxide film.[60] This protects the alloy from corrosion. For example, the addition of chromium to iron causes the oxide film to change from polycrystalline to non-crystalline as the amount of chromium increases, making it possible to produce stainless steel.[61–63]

Alloying is important such that pure Al has a high resistance to atmospheric uniform corrosion, while the aerospace alloy Al 2024, containing 5 percent Cu, among others, is very sensitive to selective aluminum leaching in aqueous environments. It is, on the other hand, less sensitive to pitting. In the case of steel, the addition of chromium as an alloying element substantially decreases the amount of pitting corrosion in addition to other corrosion types.[64]

4.2   Corrosion in Water

Second to atmospheric corrosion is corrosion in water. The rate of attack is greatest if water is soft and acidic and the corrosion products form bulky mounds on the surface as in the case of iron.[23] The areas where localized attack is occurring can seriously reduce the carrying capacity of pipes. In severe cases iron oxide can cause contamination, leading to complaints of "red water".[65] In seawater the bulk pH is 8 to 8.3; however, due to cathodic production of OH^- the pH value at the metal surface increases sufficiently for deposition of $CaCO_3$ and a small extent of $Mg(OH)_2$ together with iron hydroxides. These deposits form a surface layer that reduces oxygen diffusion. Due to this and other corrosion inhibiting compounds, such as phosphates, boric acid, organic salts, that are present, the average corrosion rate in seawater is usually less than that of soft fresh water. However, the rate is higher than it is for hard waters due their higher Ca and Mg content.[66] An exception occurs when a material is in the splash zone in seawater, where a thin water film that frequently washes away the layer of corrosion deposits exists on the surface a majority of the time, resulting in the highest oxygen supply and leading to the highest corrosion rate.[65] In slowly flowing seawater, the corrosion rate of aluminum is 1 to 5 μm/year, whereas for carbon steel it is 100 to 160 μm/year.[67] Additionally, even when the oxygen supply is limited, corrosion can occur in waters where SRB (sulfate-reducing bacteria) are active.[68] Other surface contamination, such as oil, mill scale (a surface layer of ferrous oxides of FeO and Fe_2O_3 that forms on steel or iron during hot rolling)[69] or deposits, may not increase the overall rate of corrosion, but it can lead to pitting and pinhole corrosion in the presence of aggressive anions.[70,71]

4.2.1   Cooling Water Systems

Cooling water systems are employed to expel heat from an extensive variety of applications, ranging from large power stations down to small air conditioning units associated with hospitals and office blocks.[82] Corrosion inhibitors extend the life of these systems by minimizing corrosion of heat exchange, receiving vessels and pipework that would otherwise possess a safety risk, reducing plant life and impairing process efficiency.[83] Based on the type of system present, that is, either open or closed, once-through or recirculated systems, different amounts and types of corrosion inhibitors are employed. In potable waters, for example, since the systems are non-recirculating, use of corrosion inhibitors is limited by toxicity and cost. The inhibitors used must be inexpensive and still can only be added in low quantities. Calcium carbonate, silicates, polyphosphates, phosphate and zinc salts are commonly used inhibitors in potable water. Once-through cooling waters have the similar limitation of cost. Inhibitors with sulfate, silicate, nitrite and molybdate are often used in the closed-water systems, such as steam boiler systems.[84] However, the hardness in the system may precipitate the molybdate, thus, resulting in increased inhibitor demand and corrosion of the iron material in the system.[85]

4.2.2 Oil/Petroleum Industry

In the oil/petroleum industry, corrosion of steel and other metals is a common problem in gas and oil well equipment, in refining operations and in pipeline and storage equipment.[73–77] Production tubing that carries oil/gas up from the well has the most corrosion.[78] Petroleum has water and CO_2 in water forms carbonic acid, which in turn forms $FeCO_3$. Deposits of $FeCO_3$ are cathodic relative to steel, leading to galvanic and pitting corrosion.[79] Besides water content, the salt content is also similar to seawater, and with pressures bigger than 2 bars, oil and gasses become corrosive.[80] High flow rates, high flow temperatures and the H_2S ratio in petroleum are other major factors causing corrosion.[81]

4.2.3 Mine Waters

Mine waters occupy a special place in corrosion studies considering their widely varying composition from mine to mine. Because of its low cost, availability and ease of fabrication, mild steel is widely used as a structural material in mining equipment, although it can experience rapid and catastrophic corrosion failure when in contact with polluted acid mine waters. Specifically in coal mines, corrosion is known to be a serious problem.[86]

4.3 Corrosion in Soil

Particle size of soils is an important factor on corrosion in addition to the apparent effect of acidity levels. Gravel contains the coarsest and clay contains the finest particles, with a 2 mm. diameter for the former and a 0.002 mm. diameter for the latter. Sizes of sand and silt are in between gravel and clay. While clay prevents the supply of oxygen but not water, gravels allow oxygen supply as well.[72]

In concrete, carbonation reduces the pH of solution and leads to general breakdown of passivity.[31]

5

Nature of Protective Metal Oxide Films

Regardless of the corrosion type, the major product of iron and steel corrosion is FeOOH, which is referred to as rust.[87] Rust can occur in 4 different crystalline modifications based on the type of corrosion and the environment that the corrosion takes place: α-FeOOH (goethite), β-FeOOH (akaganeite), γ-FeOOH (lepidocrocite), and δ-FeOOH (feroxyhite).[88–89]

α-FeOOH seems to be the most stable modification of the ferric oxide hydroxides. Solubility of α-FeOOH is approximately 10^5 times lower than that of γ-FeOOH. The relative amounts of α-FeOOH and γ-FeOOH depend on the type of atmosphere and the length of exposure.[89] In freshly formed rust in SO_2 polluted atmospheres γ-FeOOH is usually slightly dominant. On prolonged exposure the ratio of γ-FeOOH to α-FeOOH decreases.[90] Also in weakly acidic conditions in general γ-FeOOH is transformed into α-FeOOH depending on the sulfate concentration and temperature.[91] In marine atmospheres, where the surface electrolyte contains chlorides, β-FeOOH is found. β-FeOOH has been shown to contain up to 5% chloride ions by weight in marine locations.[92] δ-FeOOH has not been reported in rust created under atmospheric conditions on carbon steel.[93] Magnetite, Fe_3O_4, may form by oxidation of $Fe(OH)_2$ or intermediate ferrous-ferric species such as green-rust.[94] It may also be formed by reduction of FeOOH in the presence of a limited oxygen supply according to[95]

$$8FeOOH + Fe \longrightarrow 3Fe_3O_4 + 4H_2O \qquad (5.1)$$

The rust layer formed on unalloyed steel generally consists of two regions: an inner region, next to the steel/rust interface often consisting primarily of dense, amorphous FeOOH with some crystalline Fe_3O_4; and an outer region consisting of loose crystalline α-FeOOH and γ-FeOOH.[37–38, 96]

Aluminum initially forms a few nm thick layer of aluminum oxide, mainly γ-Al_2O_3 (boehmite), which in humidified air is covered by aluminum oxyhydroxide, γ-AlOOH due to hydrolysis, resulting in a double-layer structure.[40–42] Related reactions that occur within the passive film when in contact with humidity or water are as follows;

$$Al^{3+} + 3OH^- \longrightarrow AlOOH + H_2O \qquad (5.2)$$

$$Al_2O_3 + H_2O \longrightarrow 2AlOOH \qquad (5.3)$$

$$AlOOH + H_2O \longrightarrow Al(OH)_3 \qquad (5.4)$$

The probable composition of the outer layer is a mixture of Al_2O_3 and hydrated Al_2O_3, mostly in the form of amorphous $Al(OH)_3$ or α-$Al(OH)_3$ (bayerite). This outer coating of AlOOH-$Al(OH)_3$ is colloidal and porous with poor corrosion resistance and cohesive properties. The inner layer on the other hand is mostly composed of Al_2O_3 and small amounts of hydrated aluminum oxide mostly in the form of AlOOH. This inner coating of Al_2O_3-AlOOH is continuous, resistant to corrosion and is a good base for paints and lacquers.[43-45] Altogether, this passive layer is insoluble in the pH interval of 4 to 9.[46] Lower pH values results in the dissolution of Al^{3+}.[97]

6

Effect of Aggressive Anions on Corrosion

Both weight loss and salt-fog chamber tests are commonly performed under circumstances where high salt concentrations are present. For weight loss tests, high salt concentrations are applied for accelerated corrosion testing purposes in addition to simulating the actual highly corrosive environments, such as marine environments, seawater and industrial areas. In the case of salt-fog chamber tests, chemical stress in accelerated testing primarily refers to chloride containing salts in solution because airborne contaminants are believed to play a very minor role in paint aging.[461] Other chemical stress factors, such as UV effects, are not of focus here since any coating, such as a sol-gel coating, can be protected from UV exposure by simply painting over it with a paint that does not transmit light.

Many mechanisms have been proposed for the suppression or acceleration of metallic dissolution by the action of aggressive anions in general.[462,463] The simple most common theory on the accelerated corrosion due to aggressive anions is the concept of competitive adsorption. Aggressive anions, such as Cl^-, compete with adsorption of OH^- or the inhibitor ion depending on pH. Thus, aggressive anions increase the concentrations of inhibitors required to prevent corrosion. This must be taken into account; since the application of less than the adequate inhibitor concentration leads to pitting corrosion.[81] Competitive adsorption of aggressive anions can lead to corrosion in two different ways. Cl^-, for instance, may either cause the initial local breakdown of the passive oxide film or simply interfere with the repassivation process after the film has been broken down locally. In one study, no indication was found that Cl^- is incorporated into the anodic film on iron when the passive oxide film was initially formed in a Cl^- containing solution suggesting that Cl^- ions cause local film thinning by interfering with the film repair.[464-466]

In the case of aluminum adsorbed aggressive anions such as chloride can undergo a chemical reaction with the passive film and produce soluble transient compounds such as $Al(OH)_2Cl$, $AlOHCl_2$, and $AlOCl$, which are easily dissolved into the solution once they are formed.[12] Similarly, soluble $FeSO_4$ complex forms in presence of another aggressive anion, that is SO_4^{2-}.[10] Thus as a result of these adsorption-dissolution processes, the protective oxide film is thinned locally, small pits are made and the corrosion rate of aluminum is greatly enhanced.[98-100]

When aggressive anions have to be compared with one another, the stability of the intermediate complexes of substrate metal and aggressive anions must be considered. In the specific case of steel corrosion, if an anion, X^-, is first adsorbed on the steel surface, a surface complex forms in the anodic process, and then the complex is desorbed from the surface.[11, 467]

$$Fe + X^- \longleftrightarrow (Fe\,X^-)_s \qquad (6.1)$$

$$(FeX^-)_s \longleftrightarrow (Fe\,X)_s + e^- \qquad (6.2)$$

$$(FeX)_s \longrightarrow FeX^+ + e^- \qquad (6.3)$$

$$FeX^+ \longleftrightarrow Fe^{2+} + X^- \qquad (6.4)$$

s represents ion or compound at the surface. In general, if the adsorbed anion or the surface complex is stable, the corrosion of steel is suppressed. Therefore, the order of tested anions in terms of the stability of the surface complex based on the corrosion rates would be $ClO_4^- > SO_4^{2-} > Cl^-$.[467]

Due to the stability of intermediate complexes between the metal substrate and the aggressive anions, pitting corrosion does not occur for chromium metal. Stability constants of CrX^{2+} complexes are smaller than 1, for instance it is 1 when X is Cl^- and 10^{-5} when it is I^-.[8] In addition, exchange of Cl^- and H_2O ligands between the inner and outer sphere of chromium halide complexes is extremely slow.[8] Together these factors causes insolubility of $CrCl_3$ in cold water due to very low dissolution rate of Cr^{3+}. Therefore the presence of a Cr-Cl complex at the surface will not increase the dissolution rate because it will dissolve very slowly by itself. In the case of Fe^{3+} this exchange is very rapid. Similarly Fe-Cr alloys are more resistant to pitting in Cl^- solution than is pure Fe.

7

Corrosion Prevention Methods

With such variety in types of corrosion come many different prevention methods. Among these is selecting a material which does not corrode in the actual environment. When changing the material is not possible, changing the environment to prevent transport of essential reactants of corrosion often using corrosion inhibitors seems to be the second most reasonable prevention method. Using chemical inhibitors to lower molecular oxygen activity at the metal surface is one example of this type of prevention technique. Also, applying coatings on the metal surface in the form of paint, providing a barrier between the metal surface and the corrosive environment, is another very commonly used prevention technique. Other prevention techniques include, but are not limited to, using special designs to prevent water accumulation on the metal surfaces or changing the potential, which results in a more negative metal and thus prevents transfer of positive metal ions from the metal to the environment.[101]

Development of novel chemical inhibitors for mild steel and aluminum alloys constitutes the major part of research on chromate replacements. Mild steel alloy finds extensive use in various structural applications due to its physical characteristics, such as stiffness and high strength-to weight-ratios, while aluminum and aluminum alloys are widely used in engineering applications because of their combination of lightness with strength, high corrosion resistance, thermal and electrical conductivity, heat and light reflectivity and hygienic and non¬toxic qualities.[102] In addition to its mechanical properties, the low residual radioactivity is another unique property of aluminum, leading to its use as the first wall in thermonuclear reactors. However, the long and safe exploitation of aluminum alloys in nuclear power production greatly depends on its corrosion stability, which is why the type of the alloy and corrosion protection measures are important.[103]

8

Commonly Used Alloys and their Properties

The composition of alloying elements of mild steel is commonly 0.02 to 0.03 percent sulfur, 0.03 to 0.08 percent phosphorus, 0.4 to 0.5 percent manganese, and 0.1 to 0.2 percent carbon.

The aluminum alloys are usually divided into two major groups: cast alloys and wrought alloys. While the term "wrought aluminum" may not be as familiar as wrought iron, it basically refers to aluminum material that is constructed using wrought iron techniques. Essentially, this means that the aluminum is "shaped" to produce the desired material. The term "wrought iron" is slightly ambiguous, as it refers not only to the method of construction but also to the type of metal used. In other words, wrought iron is a specific type of iron and also a style of metal work, while wrought aluminum simply refers to the metalworking method, not the type of aluminum. Cast aluminum, on the other hand, is made from literally pouring molten aluminum into a cast and allowing it to harden. Each wrought and cast aluminum alloy is designated by a four-digit number by the Aluminum Association of the United States[104,105] with slight differences between wrought and cast alloys (See Table 8.1). The first digit indicates the alloy group according to the major alloying element. The second digit indicates the modification of the alloy or impurity limits. Original (basic) alloy is designated by "0" as the second digit. Numbers 1 through 9 indicate various alloy modifications with slight differences in the compositions.

The last two digits identify the aluminum alloy or indicate the alloy purity. In the alloys of the 1xxx series, the last two digits

Table 8.1 Designations for Alloyed Wrought and Cast Aluminum Alloys.

Wrought Alloy		Cast Alloy	
Name	Major Alloying Element	Name	Major Alloying Element
1xxx	More than 99% pure Al	1xx.x	More than 99% pure Al
2xxx	Cu, small amount of Mg	2xx.x	Cu
3xxx	Mn	3xx.x	Si with Cu and/ or Mg
4xxx	Si	4xx.x	Si
5xxx	Mg	5xx.x	Mg
6xxx	Mg, Si	6xx.x	Unused
7xxx	Zn, small amount of Cu, Mg, Cr, Zr	7xx.x	Zn with Cu and/ or Mg
8xxx	Other elements (Li, Ni)	8xx.x	Sn

indicate the level of purity of the alloy: 1070 or 1170 means minimum 99.70 percent of aluminum in the alloys, 1050 or 1250 means 99.50 percent of aluminum in the alloys, 1100 or 1200 means a minimum 99.00 percent of aluminum in the alloys. In all other groups of aluminum alloys (2xxx through 8xxx) the last two digits signify different alloys in the group.

8.1   Aluminum 2024 Alloy

The 2xxx (aluminum-copper) alloy series started to be used frequently with the development of 24S (2024) in 1933 for maximum solubility of alloying elements in the solid phase. Due to their high strength, toughness and fatigue resistance, modifications of 24S are widely used today for aircraft applications.[106] However, the alloys of these series, in which the copper is major alloying element, are less corrosion-resistant than the alloys of other series. Copper increases the efficiency of the cathodic counter reaction of the corrosion, such as O_2 and H^+, reduction reaction and, thus, the presence of copper increases the corrosion rate.[107]

Despite its inferior corrosion resistant properties, Al 2024 is substantially used due to the fact that it is a peculiar alloy used in the fuselage structures of aircrafts, where the corrosion resistance properties are compromised for the sake of mechanical strength also due to the characteristics of its potential environmentally friendly binders, for instance sol-gel coating.

The nominal composition of Al 2024-T3 alloy is 4.4 percent Cu, 1.5 percent Mg, 0.6 percent Mn, and lesser amounts of Fe, Si and impurity element allowable.[109–111] The "T3" designation indicates that the alloy was solution-annealed, quenched and aged at ambient temperatures to a substantially stable condition.[112]

It is important to recognize that in most modern aircraft an "alclad" variant of the 2024-T3 is used. Alclad 2024-T3 has a thin layer of commercially pure Al applied to enhance corrosion resistance.[25]

However, alclad layer is easily removed, exposing the underlying 2024T3 core in maintenance operations where the grinding out of cosmetic corrosion surfaces is routine. Thus, corrosion protection of the Al 2024T3 core then becomes an issue, especially for older aircraft that have experienced many depot maintenance cycles.[113]

8.2   Aluminum 7075 Alloy

Alloy 75S (7075), developed during World War II, provided the high-strength capability not available with aluminum-magnesium-copper alloys. This type of alloy contains major additions of Zn, along with Mg or both Mg and Cu. The Cu containing alloys have the highest strength and, therefore, have been used as construction materials, especially in aircraft applications. The Cu-free alloys, which have good workability, weldability as well as moderate strength, have increased in their applications in automotive industry.[107] The first commercial aluminum-magnesium-silicon alloy (51S) was developed and brought to market by 1921.

8.3   Aluminum 6061 Alloy

The introduction of alloy 61S (6061) in 1935 filled the need for medium-strength, heat-treatable products with good corrosion resistance that could be welded or anodized. The corrosion resistance of alloy 6061 even after welding made it popular in early railroad and marine applications. Alloy (62S) 6062, a low-chromium version of similar magnesium and silicon, was introduced in 1947 to provide finer grain size in some cold-worked products. Unlike the harder aluminum-copper alloys, this 61S and 62S alloy series of Al-Mg-Si could be easily fabricated by extrusion, rolling or forging. These alloys' mechanical properties were adequate (mid-4045 ksi range) even with a less-than-optimum quench, enabling them to replace mild steel in many markets. The moderate high strength and very good corrosion-resistant properties of this alloy series of Al-Mg-Si make it highly suitable in various structural building, marine and machinery applications. The ease of hot working and low-quench sensitivity are advantages in forged automotive and truck wheels. Also made from alloy 6061 are structural sheet and tooling plate produced for the flat-rolled products market, extruded structural shapes, rod and bar, tubing and automotive drive shafts.[108]

Detailed composition of certain aluminum alloys is given in Table 8.2;

Table 8.2 Chemical Composition of Aluminum Alloys.

Alloying Element	2024	6061	7075
Al	91.5–92.8	96.8–97.2	86.85–89.55
Cu	3.8–4.9	0.15–0.4	1.2–2.0
Mg	1.2–1.8	0.8–1.2	2.1–2.9
Mn	0.3–0.9	≤ 0.15	≤ 0.30
Fe	≤ 0.50	≤ 0.7	≤ 0.50
Si	≤ 0.50	0.4–0.8	≤ 0.40
Zn	≤ 0.25	≤ 0.25	5.1 6.1
Zr+Ti	≤ 0.20	–	≤ 0.25
Ti	≤ 0.15	≤ 0.15	≤ 0.20
Cr	≤ 0.10	0.04–0.35	0.18–0.28

9

Cost of Corrosion and Use of Corrosion Inhibitors

In a study entitled "Corrosion Costs and Preventive Strategies in the United States," conducted from 1999 to 2001 by CC Technologies Laboratories, the total annual estimated direct cost of corrosion in the United States was estimated a staggering $276 billion equaling to approximately 3.1 percent of the nation's Gross Domestic Product (GDP).[114] This cost includes the application of protective coatings (paint, surface treatment, etc.), inspection and repair of corroded surfaces and structures and disposal of hazardous waste materials. The study reveals that, although corrosion management has improved over the past several decades, the United States must find more and better ways to encourage support and implement optimal corrosion control practices. Due to reasons such as economics and ease of application, corrosion inhibitors continue to be the most common corrosion prevention technique. Compared to other techniques, corrosion inhibitors are very convenient since they can be employed alone or within a protective coating, such as paint. Also, among many developed corrosion inhibitors, it is possible to find a working one for any specific demand.[115]

The definition of corrosion inhibitor favored by the National Association of Corrosion Engineers (NACE) is "a substance which retards corrosion when added to an environment in small concentrations."[116] Alternatively, according to the American Society for Testing and Materials' corrosion glossary, a corrosion inhibitor is defined as a chemical substance or combination of substances that, when present in the proper concentration and forms in the environment, prevents or reduces corrosion.[1]

Available references in corrosion phenomena in the technical literature appeared by the end of the 18th century. The first patent in corrosion inhibition was given to Baldwin, British patent 2327.[117]

Corrosion inhibition is reversible, and a minimum concentration of the inhibiting compound must be present to maintain the inhibiting surface film. Good circulation and the absence of any stagnant areas are necessary to maintain inhibitor concentration.[118]

Inhibitors function in one or more ways to control corrosion, namely by adsorption of a thin film onto the surface of a corroding material, by inducing the formation of a thick corrosion product or by changing the characteristics of the environment, resulting in reduced aggressiveness. Some remove oxygen in the aqueous media to reduce the cathodic reaction. Though there are many chemicals that can function as inhibitors, some may be too expensive and not economical. Chemicals that are toxic or not environmentally friendly are also of limited use. Moreover, inhibitors for one metal may or may not work for another or even may cause corrosion. In addition, the effectiveness of inhibitors is affected by the pH, temperature and water chemistry of the system.[119]

Generally, inhibitors efficient in acid solutions have little or no effect in near-neutral aqueous solutions, since in acidic media the main cathodic process is hydrogen evolution and inhibitor action is due to adsorption on oxide-free metal surfaces.[120] In alkaline conditions, most metals are inclined to be passive and are protected from most of the corrosion damage.[121] In near-neutral solutions, in which the cathodic half-reaction is oxygen reduction, corrosion processes result in the formation of sparingly soluble surface products, such as oxides, hydroxides and salts. Therefore, the inhibitor action must be exerted on the oxide-covered surface by increasing or maintaining the protective characteristics of the oxide or surface layers in aggressive solutions.[122–123]

10

Types of Corrosion Inhibitors

While there are various inhibitor classifications listed in the literature, there is no completely satisfactory way to categorize. One of the common ways is to classify them according to their reaction at the metal surface.[1,124] Based on this criterion:

1. Anodic inhibitors are compounds that reduce the actual rates of the metal dissolution that is the anodic reaction.
2. Cathodic inhibitors are compounds that reduce the rates of the cathodic reactions, such as the hydrogen evolution or oxygen reduction reactions.
3. Mixed inhibitors are compounds that retard the anodic and cathodic corrosion processes simultaneously by general adsorption covering the entire surface, sometimes with a polymer.

43

10.1   Anodic Inhibitors

Anodic or passivating inhibitors slow down corrosion by either stabilizing or repassivating the damaged passive film by forming insoluble compounds or by preventing adsorption of aggressive anions via competitive adsorption. They are used in the neutral pH range to treat cooling water systems, cooling system metals, and steel-concrete composites.[125] Passivating inhibitors can be further divided into two types: direct passivating inhibitors, which are oxidizers themselves, and indirect passivating inhibitors, which are nonoxidizers and require the presence of oxygen.[126] Direct passivating inhibitors react with metals directly and become incorporated into the passive film to strengthen it, complete it and repair it.[127] Chromate (CrO_4^{2-}) and nitrites (NO_2^-) are the best oxidizers that can passivate steel in deaerated solutions; however, both inhibitors have limited uses due to toxicity.[128] In open systems, oxygen is abundant enough, while in closed systems the addition of oxidizing salts is needed for indirect passivating inhibitiors (e.g. molybdates, or other analogues of chromates) to function.[129,130] Indirect passivators may develop a protective film in the form of a salt. It is proposed, for example, that ferrous ions at the solution/metal interface react with molybdate ions to form a complex which is further oxidized to an insulative ferric-molybdate and covers the metal surface with a thin, adherent protective film.[131–132]

10.2   Cathodic Inhdibitors

Cathodic Inhibitors slow down corrosion by reducing the rate of the cathodic reaction in the corrosion system. They may form precipitates in the cathodic locations to limit access of the cathodic reaction species, and they are also called precipitation inhibitors.[133] Zinc salts are cathodic inhibitors that form precipitates of zinc hydroxide at the cathode.[134] Magnesium salts also work in a similar way.[135] Bicarbonate (HCO_3^-) forms insoluble metal carbonates in alkaline solution.[136] Phosphates, the most widely used corrosion inhibitors of steel, precipitate as ferrous and ferric phosphates on the substrate surface.[137] Oxygen scavengers react with the dissolved oxygen to limit the supply of oxygen for the cathodic reaction. Sodium sulfite is an oxygen scavenger commonly used at room temperatures. It reacts with oxygen to form sulfate. However, since oxygen scavengers remove oxygen only, they are not effective in acidic media.[138] Cathodic poisons make discharges of hydrogen gas difficult.[139] Cathodic inhibitors are generally not as effective as anodic inhibitors (passivators), but, on the other hand, they are not likely to cause pitting.[140]

As for organic inhibitors, chelating agents, which contain at least two functional polar groups, such as acidic –COOH, –SH or basic –NH$_2$ groups, those able to form coordinate bonds with metal cations are good examples.[141] Gluconate is a complexing agent with two carboxylic groups.

11

Chromates: Best Corrosion Inhibitors to Date

Overall, chromates as inhibitors and in chromate conversion coatings as protective coatings continue to be the most efficient corrosion prevention methods for the most commonly used metals, such as steel, aluminum, zinc and magnesium among others.[142] The term conversion coating here refers to the traditional surface passivation treatment for steel and aluminum, which produces a layer of corrosion product by means of dissolution of the base metal through reaction with the passivating solution and precipitation of insoluble compounds, capable of resisting further chemical attack.[115,143] Chromate conversion coatings used for aluminum, typically generated from mixtures of soluble hexavalent chromium salts and chromic acid, participate in oxidation-reduction reactions with aluminum surfaces,[144] precipitating a continuous layer of insoluble trivalent compounds.[145] The use of chromate conversion coatings to increase the corrosion resistance and paintability of aluminum alloys can be traced to the early part of the 20th century.[146] The protection of many aluminum alloys, such as those used in aerospace components, depends heavily on chromates. Of particular interest to the Navy is the use of chromate conversion coatings on aircraft aluminum alloys, owing to excellent corrosion resistance and the ability to serve as an effective base for paint.[147–149]

Only films formed in chromate solutions meet the stringent corrosion resistance requirements of the military specifications MILC81706.[150] It is estimated that about 100,000 tonnes of aluminum per year in the U.K. are chromate treated. An anodized film may be substituted for chromate conversion coatings on certain aluminum products but only at greater operating and capital costs.[97]

Among advantages of the chromate conversion coatings are good paint adhesion, low cost, quick and simple application process by immersion, spray, rolling, the capability to resist forming operations and excellent corrosion resistance, including a self-healing ability.[151]

Results from exposure corrosion testing show that aluminum surfaces prepared with a chromate conversion coating and a chromate-free primer perform much better than a chromate-free sol-gel type of conversion coating with the same chromate-free primer,[152] leading to the necessity for enriching the sol-gel coating with efficient inhibitors.

11.1   Limitations on the Use of Chromates due to Toxicity

The mobility of aqueous Cr^{6+} within biological systems and its reactivity with biochemical oxidation mediators make it highly toxic, carcinogenic and generally regarded as a very hazardous soil and groundwater pollutant.[102, 143, 153–156]

More rigid environmental regulations have been introduced about the use of chromates, mandating the elimination of hexavalent chromium as the active ingredient in corrosion inhibition packages for the protection of aluminum-skinned aircraft.[157–158] The harmful effects of chromates on human tissue have been well documented. Dermatitis and skin cancer have been reported among workers merely handling components protected by a chromate film.[97] Many reviews in the literature points out to toxicity of chromates, such an association of Cr^{6+} with lung cancer. Although there is no general agreement on the details for the Cr^{6+} induced damage to DNA resulting in cancers, it is clear that Cr^{6+} is highly water soluble and it passes through cell membranes, and highly reactive intermediates such as Cr^{5+} stabilized by alpha hydroxyl carboxylates and Cr^{4+} are genotoxic and react either directly or through free radical intermediates to damage DNA.[159–164] Also, adverse toxicity of chromates to aquatic life has always been a problem. Chromate is quoted on the EU Red List of the EU Dangerous Substances Directive No 76/464/EEC and Groundwater Directive No 80/68/EEC.[81]

National Primary Drinking Water Regulations prepared by EPA (Environmental Protection Agency) states that chromium is a naturally occurring element found as chrome iron ore, primarily as chromite ($FeO.Cr_2O_3$), in rocks, animals, plants, soil, and in volcanic dust and gases.[165–168] In air, chromium compounds are present mostly as fine dust particles, which eventually settle over land and water. Chromium can strongly attach to soil and only a small amount can dissolve in water and move deeper in the soil to underground water. There is also a high potential for accumulation of chromium in aquatic life.[165,167]

Chromium is present in the environment in several different forms. The most common forms are Cr(0), Cr(III) and Cr(VI). No taste or odor is associated with chromium compounds. Cr(III) occurs naturally in the environment and is an essential nutrient. Cr(VI) and Cr(0) are generally produced by industrial processes. The metal chromium, which is the Cr(0) form, is used for making steel. Cr(VI) and Cr(III) are used for chrome plating, dyes and pigments, leather tanning by means of chromic sulfate, wood preserving by means of copper dichromate, treating cooling tower water, magnetic tapes, cement, paper, rubber, composition floor covering, automobile brake lining and catalytic converters and other materials. Smaller amounts are used in drilling muds, textiles and toner for copying machines.[165–168] Production of the most water-soluble forms of chromium, the chromate and dichromates, was in the range of 250,000 tons in 1992.[165,167] The two largest sources of chromium emission in the atmosphere are from the chemical manufacturing industry and combustion of natural gas, oil and coal. The following treatment methods have been approved by the EPA for removing chromium: coagulation/filtration, ion exchange, reverse osmosis and lime softening.[165] From 1987 to 1993, according to the Toxics Release Inventory, chromium compound releases to land and water totaled nearly 200 million pounds. These releases were primarily from industrial organic chemical industries. The largest releases occurred in Texas and North Carolina. The largest direct releases to water occurred in Georgia and Pennsylvania. In 1974, Congress passed the Safe Drinking Water Act Law, which requires the EPA to determine safe levels of chemicals in drinking water that do or may cause health problems.[165,167] The Maximum Contaminant Level Goal (MCLG) for chromium has been set at 0.1 parts per million (ppm), because the EPA believes this level of protection would not cause any of the potential health problems described below. Based on this MCLG, the EPA has set an enforceable standard called a Maximum Contaminant Level (MCL). MCLs are set as close to the MCLGs as possible, considering the ability of public water systems to detect and remove contaminants using suitable treatment technologies. The MCL has also been set at 0.1 ppm because the EPA believes, given present technology and resources, this is the lowest level to which water systems can reasonably be required to remove this contaminant should it occur in drinking water. The Reference Concentration (RfC) for Cr(VI) (particulates) is 0.0001 mg/m^3 based on respiratory effects in rats. The RfC for Cr(VI) (chromic acid mists and dissolved Cr(VI) aerosols) is 0.000008 mg/m3 based on respiratory effects in humans. The

EPA has not established an RfC for Cr(III). The RfD for Cr(III) is 1.5 mg/kg/d based on the exposure level at which no effects were observed in rats exposed to Cr(III) in the diet.[165–168]

The general population is exposed to chromate by eating food, drinking water and inhaling air that contains the chemical. The average daily intake of chromium, generally in the form of Cr(III), from air, water, and food is estimated to be less than 0.2 to 0.4 micrograms (µg) from air, 2.0 µg from water, and 60 µg from food, respectively.[166,168]

The EPA reports hexavalent chromium to cause shortness of breath, coughing, wheezing (mostly with inhalation of chromium trioxide) and skin irritation or ulceration, when people are exposed to it at levels above the MCL for relatively short periods of time, while damage to circulatory and nerve tissues, stomach upsets and ulcers, convulsions, kidney and liver damage, perforations and ulcerations of the septum, bronchitis, asthma, decreased pulmonary function, pneumonia, skin irritation and even death are potential results of a long-term or a lifetime exposure. Some people are extremely sensitive to Cr(VI) or Cr(III). Allergic reactions consisting of severe redness and swelling of the skin have been noted. Long-term exposure to Cr(VI) has been associated with lung cancer, as in the case of workers exposed to levels in air that were 100 to 1,000 times higher than those found in the natural environment. Lung cancer may occur long after exposure to chromium has ended. Limited information on the reproductive effects of Cr(VI) in humans exposed by inhalation suggest that exposure to Cr(VI) may result in complications during pregnancy and childbirth.[165,167]

On the contrary, Cr(III) is an essential nutrient, with a daily intake of 50 to 200 µg recommended for adults. This ion helps the body use sugar, protein, and fat. Without Cr(III) in the diet, the body loses its ability to use sugars, proteins and fat properly, which may result in weight loss or decreased growth, improper function of the nervous system and a diabetic-like condition. With too much intake, Cr(III) can also cause health problems, but it is considered about 100 to 1,000 times less toxic than Cr(VI). Although each form can be converted to the other form under certain conditions, Cr(III) is not oxidized to Cr(VI) in the natural soil environment.[166,168]

Cr(III) compounds are one of the major candidates to replace Cr(VI), based corrosion inhibitors and protective coatings if the required corrosion resistance and adhesion of organic coatings can be obtained.[153] Thus, Cr(III) compounds were investigated in this project as chromate replacements. Cr(III) is not an oxidizing agent, but it will form the mixed oxides/ hydroxides with the substrate in the presence of a primary passivator/oxidizing agent, such as dissolved oxygen. When a primary oxidizing agent is present, the substrate can oxidize to its higher oxidation state cations, producing hydroxide, and the existing Cr(III) ions can react with the produced hydroxides to form a conversion coating composed of mixed oxides/ hydroxides of the substrate and Cr(III).[97]

The metal Cr(0) is less common and does not occur naturally. It is not clear how much it affects health, but it is not currently believed to cause a serious health risk.[169]

The International Agency for Research on Cancer (IARC) has determined that Cr(VI) is carcinogenic to humans. IARC has also determined that Cr(0) and Cr(III) compounds are not classifiable as to their carcinogenicity to humans.[170,171] The World Health Organization (WHO) has determined that Cr(VI) is a human carcinogen.[171] The Department of Health and Human Services (DHHS) has determined that certain Cr(VI) compounds (calcium chromate, chromium trioxide, lead chromate, strontium chromate, and zinc chromate) are known human carcinogens.[172] Finally, the EPA has classified Cr(VI) as a Group A, known human carcinogen by the inhalation route of exposure.[165–168,173–176]

In the light of given negative effects of hexavalent chromium compounds, stricter environmental regulations have already mandated their removal from water and general waste effluents and have mandated their near-term removal from corrosion inhibiting packages used for the protection of aluminum-skinned aircraft.[149,157,177–180]

Strict regulations already exist for chromate residues that require the use of expensive effluent treatments to achieve the desired residual concentrations by precipitating hexavalent chromium compounds.[97,181] Despite their negative aspects, to date, no replacements exist in the market for carcinogenic chromates with the same efficiency for a range of aluminum alloys and steel, neither as pigment nor as a metal pretreatment.[110,182]

For perhaps the last 20 years or more, a considerable effort has focused on discovering nonchromate corrosion-inhibiting compounds for protection of aluminum alloys. A number of reviews focusing on this subject alone have been written in the past several years.[183,184,185]

Given the toxicity and carcinogenicity of chromates, the purpose shall not be only to synthesize efficient corrosion inhibitors for certain alloys of certain metals to be applied in different environments, but also to find environmentally friendly corrosion inhibitors for successful chromate replacements. In this regard, the standard for an environmentally friendly inhibitor is considered as having acceptable or no toxicity compared to chromate inhibitors. Studying the reasons underlying the success of chromate inhibitors seems to be the first reasonable approach one might take before formulating chromate replacements.

11.2   Corrosion Inhibition Mechanism of Chromates

Chromates are very effective inhibitors of Fe, Al, Cu, Zn corrosion. The unique chemical and electronic properties of the oxo-compounds of chromium give rise to a unique ability to inhibit corrosion in ferrous and nonferrous materials.[186] They are both anodic and cathodic inhibitors due to their abilities to form precipitates with the dissolving metal ions such as iron, aluminum, and zinc ions, at anodic sites and by reducing to trivalent chromium to form composite inert compounds at cathodic sites.[187] The tetrahedral, d^0, hexavalent Cr^{6+} oxoanion compounds of chromium, which are chromate, dichromate, bichromate, and chromic acid, dissolve as stable and mobile complexes in water. Thus, they are easily transported to sites of localized corrosion where they are reduced to very stable, kinetically inert refractory oxide compounds of Cr^{3+}.[188] These octahedral, trivalent, d^3, compounds of Cr^{3+} are irreversibly adsorb at metal and metal oxide surfaces to form a protective film of a near monolayer thickness.[189] As one of these irreversibly adsorbed compounds, $Cr(OH)_3$ provides a good, hydrophobic barrier with good adhesion properties.[190] The concentration of the transported or leached chromate is sufficient to be active as an inhibitor for the metal under the paint, at defects or at cut edges. These hexavalent oxoanion compounds of chromium also have optimum solubilities enabling them to be used as efficient paint pigments, in which blistering of the paint does not occur.[147] Also, possibly the most crucial property of the barrier film of trivalent chromium compounds is its ability to store Cr^{6+} oxoanions that can be slowly released into a solution when attacked by aggressive anions. These released Cr^{6+} oxoanions can migrate to and interact at defects to interrupt corrosion, which gives rise to the unique "self-healing" ability of chromate conversion coatings in general. There is a good agreement that chromate conversion coatings not only contain but also release hexavalent chromium to repair defects and damage of the conversion coating.[190–199]

Specifically for aluminum corrosion; released Cr^{6+} oxoanions inhibit pit initiation by adsorbing onto aluminum oxides, thereby discouraging adsorption of anions such as chloride and sulfate, which promote dissolution and destabilization of the protective oxides.[200–201] Thus, competitive adsorption of chromates with regard to aggressive anions such as chloride and sulfate appears as another major property of chromate conversion coatings.[202]

Along with nitrites, chromates passivate independent of dissolved oxygen in contrast to molybdates and vanadates, which require the presence of dissolved oxygen as a primary passivator.[203]

In general, following steps of reactions occur:[204]

$$Cr^{6+} \longrightarrow Cr^{3+} + 3H_2O \longrightarrow Cr(OH)_3 + 3H^+ \longrightarrow Cr_2O_3 \cdot 3H_2O \quad (11.1)$$

The hydrolysis reactions generate H^+, which are consumed by redox reactions. In alkaline conditions,[205]

$$CrO_4^{2-} + 4H_2O + 3e^- \longrightarrow Cr(OH)_3 + 5OH^- \quad (11.2)$$

In case of iron corrosion in near neutral conditions,[206]

$$6FeO + 2CrO_4^{2-} + 2H_2O \longrightarrow Cr_2O_3 + 2Fe_2O_3 + 4OH^- \quad (11.3)$$

Mixed chromium/iron hydroxides also form such as,[207]

$$3Fe^{2+} + HCrO_4^{2-} + 8H_2O \longrightarrow Fe_3Cr(OH)_{12} + 5H^+ \quad (11.4)$$

In contrast to nitrites, molybdates, vanadates and other inhibitors, chromates are also effective in moderately acidic conditions. In an acidic medium, CrO_4^{2-} converts to $Cr_2O_7^{2-}$, which is a very strong oxidant, according to

$$Cr_2O_7^{2-} + 14H^+ + 6e^- \longrightarrow 2Cr^{3+} + 7H_2O,[208] \quad (11.5)$$

The following reaction takes place with the metal substrates[209]

$$6/n\, M^0 + Cr_2O_7^{2-} + 7H^+ \longrightarrow 6/n\, M^{n+} + 2Cr^{3+} + 7H_2O \quad (11.6)$$

where M^0 can be Al, Fe, Zn.

For the specific case of chromium conversion coating formation on Al, the following overall formation reactions are given;[210, 211]

$$Cr_2O_7^{2-} + 2Al + 2H^+ + H_2O \longrightarrow CrOOH + 2AlOOH \quad (11.7)$$

or[212]

$$Cr_2O_7^{2-} + 2Al + 2H^+ + 2H_2O \longrightarrow 2Cr(OH)_3 + Al_2O_3 \quad (11.8)$$

The chromate conversion coating process is aided by fluoride, which prevents rapid passivation of the Al surface, thus allowing Cr^{6+} to Cr^{3+} reduction and is also aided by ferricyanide, which functions as a mediator between Al oxidation and chromate reduction and accelerates the redox reaction.[213]

As a result of these multiple redox reactions, while hexavalent Cr(VI) is reduced to its lower oxidation state oxides and hydroxides, the substrate metal is oxidized to its oxides and hydroxides. The pH also rises to the point where trivalent chromium and other oxide/hydroxide compounds are insoluble.[214] Consequently, a protective conversion coating of adherent composite oxide/hydroxides[215] form with the general formula of M_2O_3/Cr_2O_3 and/or $M(OH)_3/Cr(OH)^3$, where $M^0 = Fe, Al$.[216]

Another reason for protective ability of chromium oxide and hydroxide film over aluminum surfaces is their stability over a wider range of pH. Based on Pourbaix-diagrams, the approximate stability limit of the Al oxide is at pH 9, while it is up to pH 15 for Cr(III) oxide.[217]

12

Chromate Inhibitor Replacements: Current and Potential Applications

Given some basic information about the corrosion inhibition mechanisms of chromates, many studies have been conducted for chromate replacements. For effective replacement of hexavalent Cr, however, an inhibitor has to inhibit the oxygen reduction reaction as well as anodic dissolution/pitting, and several studies indicate that hybrid formulations seem to be the best way to do just that. Typically, in these hybrid formulations an organic oxygen reduction reaction inhibitor is included with environmentally benign anodic inhibiting anions.

12.1 Nitrites

Other commonly used inhibitors that passivate independent of dissolved oxygen are nitrites. Nitrites are the established inhibitors for rusting machinery tooling and workpieces, and they are often used with alkanolamines. However, like chromates, they are also being replaced because of the risk of carcinogenic nitrosamine formation.[218]

Nitrites' Maximum Contaminant Level (MCL) and Maximum Contaminant Level Goal (MCLG) limits have been determined as 1 mg/L each by the Environmental Protection Agency (EPA). Infants below the age of six who drink water containing nitrite in excess of the MCL could become seriously ill and, if untreated, may die. Symptoms include shortness of breath and blue baby syndrome.

Major nitrite sources are listed as runoffs from fertilizer uses, leaches from septic tanks and sewages.[219]

12.2 Trivalent Chromium Compounds

Cr(III) compounds arise as one of the potential replacements for Cr(VI) compounds given its much lower toxicity. Cr(III) is not an oxidizing agent but it will form the mixed oxides/hydroxides with the substrate. Therefore, in the presence of a primary passivator/oxidizing agent, such as dissolved oxygen, the substrate can oxidize to its higher oxidation state cations, producing hydroxide and the existing Cr(III) ions would react with the produced hydroxides to form a conversion coating composed of mixed oxides/hydroxides of the substrate and Cr(III).[149,220–221] Despite this, there are limited successful applications of trivalent chromium coatings. The corrosion resistance of trivalent chromium coatings was found considerably less effective than that of hexavalent Cr conversion coatings, as significant concentrations of localized pitting were observed after a 168 hr. salt spray test.[179] Thus, rather than using trivalent Cr coatings alone, incorporation of corrosion inhibitors based on trivalent Cr compounds into coatings that have better mechanical properties seems to be a more reasonable prevention method.

Formation of trivalent Cr hydroxides is based on their ability to form coordination compounds of coordination number six.

The hydrolysis of coordination complexes is accelerated by addition of alkali and the hydroxides may form successively in the following:[222]

$$[Cr(H_2O)_6]Cl_3 \longrightarrow [Cr(OH)(H_2O)_5]Cl_2 + HCl \qquad (12.1)$$

$$[Cr(OH)(H_2O)_5]Cl_2 \longrightarrow [Cr(OH)_2(H_2O)_4]Cl + HCl \qquad (12.2)$$

$$[Cr(OH)_2(H_2O)_4]Cl \longrightarrow [Cr(OH)_3(H_2O)_3] + HCl \qquad (12.3)$$

These species can polymerize as shown in eq. 12.4

$$2[Cr(OH)(H_2O)_5]Cl_2 \longrightarrow [(H_2O)_4Cr \overset{\overset{\displaystyle H}{\displaystyle O}}{\underset{\underset{\displaystyle H}{\displaystyle O}}{\diagdown \diagup}} Cr(H_2O)_4]Cl_4 + 2H_2O \qquad (12.4)$$

One concern is that Cr(III) and Al(III) compounds are both capable of forming octahedral complexes, and the introduction of these ions into an aqueous electrolyte will interfere with conversion of the hydrous alumina into the aluminum hydroxide film by bonding to the active film sites. Therefore, similar to their application in hexavalent chromate conversion coatings, fluoride ions are used to remove aluminum oxide and hydroxide films on the substrate surface before forming trivalent chromium conversion coatings.[223]

$$Al_2O_3 + 12F^- + 3H_2O \longrightarrow 2AlF6^{3-} + 6OH^- \qquad (12.5)$$

12.3 Oxyanions Analogous to Chromate

Other likely candidates to replace chromates are reducible hypervalent transition metals similar to chromium, which are compounds of Mo, V, Mn, and Tc. The high-valent oxoanions of these elements exist in aqueous solution, and they reduce to form insoluble oxides, which exhibit high resistance to dissolution in an alkaline environment in the same way as Cr.[224–227] Other anodic inhibitors might include oxo-compounds of P and B as well.[228] Among these analogous metals, however, the oxoanion of hypervalent Mn, permanganate, is thermodynamically unstable with respect to the oxidation of water unless the solution is sufficiently alkaline and all technecium isotopes are radioactive.[229–230] Vanadium oxide is relatively more stable toward high pH and Mo oxide is stable toward lower pH values.[231] Solely as oxides, the elements of Mo and V will never give the same stability as seen in analogous Cr^{3+} oxide.[232] On the other hand, the oxo-compounds of these elements can form very stable polyoxometallates with each other, or phosphates and tungstates, providing significant inhibition for aluminum corrosion, particularly when combined with other compounds.[233–234]

The inhibition mechanism of aluminum corrosion by molybdates, vanadates and similar oxyanions is primarily due to the competitive adsorption of these anions with aggressive anions, such as chloride and sulfate anions. As a result of adsorption of oxyanions in the place of aggressive anions, oxygen bridged complexes with the metal substrates form. Such complexes were found in catalyst systems such as $MoO_3+Al_2O_3$ and $WO_3+Al_2O_3$.[235–239] These compounds are expected to have a low solubility in the electrolyte hindering the dissolution of the passive film and retarding pit initiation and propagation of pitting corrosion.

Oxides of heavier elements, such as Nb, Hf, Ti, Zr, and Ta are very stable in their highest oxidation state. The mechanism for rare-earth inhibition seems to originate from the alkaline precipitation of protective oxide films at active cathodes. However, soluble and mobile precursors of these oxides remain difficult to stabilize in aqueous solution with the slight exception of Ce, which is the only lanthanide element that exhibits a tetravalent oxidation state that is stable as a complex in aqueous solution.[335,336] Ce^{4+} behaves somewhat like Cr^{6+}. The reduction product, Ce^{3+}, however, is not nearly as stable as compared to Cr^{3+}.[337]

12.3.1 Molybdates

The molybdates have been the most investigated metal oxyanion analogues. Although Molybdenum (Mo) compounds are not totally harmless, they are rapidly excreted by the body.[240] Unlike many other transition metals, molybdenum has been described as having an extremely low or even negligible toxicity.[241] In a review it is stated that, in spite of considerable use of molybdenum in industry, no incidences have been reported yet due to industrial poisoning by Mo.[242] Molybdenum compounds are listed in the lowest potentially carcinogenic category.[243–245]

The most recent Threshold Limit Value (TLV) published by the American Conference of Government Industrial Hygienists 1984–1985 show the time-weighted average TLV for soluble molybdenum particulates to be 5 mg/m^3 and for insoluble particulates to be 10 mg/m^3. For comparison, the TLV for total particulates in the nuisance dust category is 10 mg/m^3.[246] Molybdenum has long been identified as a micronutrient essential to plant life,[247–248] and as playing a major biochemical role in animal health as a constituent of several important enzyme systems.[249–250] Several studies have indicated that molybdenum-deficient diets may be associated with the incidence of various forms of cancer.[251–255]

From an environmental perspective, five statutes, and their associated regulations, govern the use and disposal of chemicals within the United States. These are Safe Drinking Water Act,[256] the Resource Conservation and Recovery Act (RCRA),[257] the Clean Water Act (CWA),[258] the Comprehensive Environmental Response Liability Act,[259] and the Toxic Substances Control Act (TSCA).[260] Molydenum is not a regulated parameter under any of these statutes. The TSCA requires all existing chemical substances be registered. Sodium molybdate has been assigned the

Chemical Abstract Service number of 7631-95-0, for instance, but has not been selected for toxicity testing. Under RCRA, Mo is neither listed as a hazardous waste nor a hazardous constituent. Section 311 of the CWA lists 299 substances as hazardous if spilled in waterways; no Mo compound is included. In summary, sodium molybdate and other molybdates are free of accompanying toxic elements or compounds and exhibit an environmental compatibility within the framework of their commercial utilization as a corrosion inhibitor.

Furthermore, molybdate inhibitors are recommended by the U.K. Health and Safety Executive Guideline (HSG70) as part of a complete water treatment program designed to minimize the risk of infecting cooling systems with the pathogen Legionella Pneumophila.[81]

Molybdenum occurs naturally in various ores; the principal source being molybdenite (MoS_2). Molybdenum compounds are used primarily in the production of metal alloys. Molybdenum is also considered an essential trace element with the provisional recommended dietary intake of 75-250 µg/day for adults and older children.[261] There is no information available on the acute or subchronic oral toxicity of molybdenum in humans. Subchronic and chronic Reference Concentrations (RfC) for Mo are not available. Information on the inhalation toxicity of Mo in humans following acute and subchronic exposures is also not available. The chronic oral Reference Dose (RfD) for Mo and Mo compounds is 0.005 mg/kg/day, based on biochemical indices in humans. The subchronic RfD is also 0.005 mg/kg/day. Mo is placed in EPA Group D, not classifiable as to carcinogenicity in humans.[261]

Corrosion-inhibiting behavior was first attributed to the molybdates in 1939.[262] First they were used as pigments;[263] and in a wide variety of applications as corrosion inhibitors.[264–275,285–298] Specifically, they have been utilized in alcohol-water antifreezes to protect automobile cooling systems from corrosion since 1939.[283–284] Molybdate allows the partial[276–278], or in complex formulations, the complete replacement of nitrite.[279–281] In addition to being efficient, molybdate inhibitior replacements for nitrites and others were found to be cost-effective.[282] Typically, a Mo concentration of 50-150 ppm is maintained in the closed cooling water systems and the pH level is maintained within the range of 9.0-10.5.[299–302] Even with concentrations insufficient to produce a layer, Mo(VI) is effective in improving the barrier properties of oxide or other films.[103–303]

In addition to the general competitive adsorption of oxyanion analogues with those of aggressive anions, as in the case of chromates, the protective effect for steel by $MoO4^{2-}$ may also be due to oxygen atoms produced via the reduction of the Mo^{6+} to Mo^{4+} (or MoO_2) during film formation,

$$MoO_4^{2-} \longrightarrow MoO_2 + 2O^- \qquad (12.6)$$

These oxygen atoms interfere with the ability of Cl^- like anions to reach the metal/film interface. The formation of MoO_2 in neutral medium is predicted by the Pourbaix diagram for Mo.[304] Also, the inhibitive nature of molybdate anions may be due to the formation of a thin film of molybdate in a range of reducible valency states, resulting in a passivating effect at anodic sites on the metal surface like other oxyanion analogues of chromate.[305]

In the case of molydate assisted inhibition of aluminum corrosion, it is believed that a layer of boehmite, $Al_2O_3.H_2O$, is formed on the surface of the aluminum specimen accompanied by a closure of the cavities with the alkali molybdate that is adsorbed on the surface. The oxidation state of Mo on the aluminum surface greatly depends on the type of molybdate that is used. It is Mo^{4+} when simple MoO_4^{2-} is used, and it is Mo^{5+} when polymolybdates are used.[306] Other theories on molybdate inhibition in the literature are widely available.[307–310]

12.3.2 Vanadates

Vanadium is a metallic element that occurs in six oxidation states and numerous inorganic compounds. Some of the more important compounds are vanadium pentoxide (V_2O_5), sodium metavanadate ($NaVO_3$), sodium orthovanadate (Na_3VO_4), vanadyl sulfate ($VOSO_4$), and ammonium vanadate (NH_4VO_3). Vanadium is used primarily as an alloying agent in steels and non-ferrous metals.[364] Vanadium compounds are also used as catalysts and in chemical, ceramic or specialty applications. An inhalation reference concentration has not been derived for vanadium or its compounds.[364] There is no information available on the acute or subchronic oral toxicity of vanadium in humans. Subchronic and chronic Reference Concentrations (RfC) for vanadium are not available. Reference Doses (RfD) for chronic oral exposures are: 0.007 mg/kg/day for vanadium; 0.009 mg/kg/ day for vanadium pentoxide; 0.02 mg/kg/day for vanadyl sulfate; and 0.001 mg/kg/day for sodium metavanadate. The subchronic RfDs for these compounds are the same as the chronic RfDs, except for sodium metavanadate, which is 0.01 mg/kg/day. There is little evidence that vanadium or vanadium compounds are reproductive toxins or teratogens. There is also no evidence that any vanadium compound is carcinogenic; however, very few adequate studies are available for evaluation. Vanadium has not been classified as to carcinogenicity by the U.S. EPA.[364]

Like molybdates and other oxyanion analogues of chromates, the inhibitive action of monovanadate anions are attributed to their competitive adsorption on the metal surface, the formation of an adsorbed layer on the oxide film and the formation of a highly insoluble salt with dissolved metal ions, which prevents the penetration of Cl^- ions and consequently decreases the rate of corrosion.[81]

It is proposed that vanadates undergo a reduction to a four-valent state upon incorporation into the surface coating of aluminum similar to MnO_2, and MoO_2.[365] Therefore, the protective ability of the four-valent oxides is a pure barrier protection, while hexavalent state compounds work as passivators.

12.3.3   Salts of Polyhydroxycarboxylic Acids

Nontoxic organic chemicals that are efficient as corrosion inhibitors include sodium, calcium and zinc salts of polyhydroxycarboxylic acids. The gluconic acid derivatives were found to hinder general corrosion of carbon steel in near-neutral media.[180] Many studies have been carried out on the use and the mechanism of action of sodium, calcium, zinc and borogluconates as corrosion inhibitors for metals, particularly for carbon steel in the neutral environment.[340–357] In other studies gluconate salts were tested as nontoxic, environmentally friendly inhibitors to replace the currently used inhibitors in cooling water systems.[358–362] There are other applications of gluconates in addition to cooling waters, such as their use to improve the corrosion resistance of medical instruments in sterilizing solutions[363] and structures in marine environments.[66] Calcium and zinc gluconates are used as dietary supplements and as first-aid treatments, while iron gluconate is used for the treatment of iron deficiencies. However, zinc appears on the list of EPA as a pollutant, but the permissible content in potable water, declared by the World Health Organization (WHO) is 5.0 mg/L as opposed to 0.1 mg/L of hexavalent chromium.[366]

Since it is listed as a secondary pollutant of drinking water, some basic information about its uses and toxicity levels are produced herein. Zinc is used primarily in galvanized metals and metal alloys, but zinc compounds also have wide commercial applications as chemical intermediates, catalysts, pigments, vulcanization activators and accelerators in the rubber industry, UV stabilizers and supplements in animal feeds and fertilizers. They are also used in rayon manufacturing, smoke bombs, soldering fluxes, mordants for printing and dyeing, wood preservatives, mildew inhibitors, deodorants, antiseptics, astringents and as rodenticides.[367] Zinc is an essential element with recommended daily allowances ranging from 5 mg for infants to 15 mg for adult males. In some medical treatment it is recommended 50 mg of zinc to be taken per day as zinc gluconate.[181] The upper limit of zinc in drinking waters is given as 5 mg/L. An inhalation reference concentration has not been derived for zinc or zinc compounds. There is no information available on the acute or subchronic oral toxicity of zinc in humans. Subchronic and chronic Reference Concentrations (RfC) for zinc are not available. The currently accepted Reference Doses (RfD) for both subchronic and chronic exposures is 0.2 mg/kg/day based on clinical data demonstrating zinc-induced copper deficiency and anemia in patients taking zinc sulfate for the treatment of sickle cell anemia. No case studies or epidemiologic evidence has been presented to suggest that zinc is carcinogenic in humans by the oral or inhalation route. Zinc is placed in weight-of-evidence Group D, not classifiable as to human carcinogenicity due to inadequate evidence in humans and animals.[367]

12.4 Synergistic Use of Oxyanions Analogues of Chromate

Despite many similarities, oxyanion analogues of chromate are not strong oxidants like chromate, and only in the presence of a primary passivator can they inhibit corrosion as anodic inhibitors. Therefore, their combined use with those of synergistic constituents in formulations seems to be a reasonable approach for obtaining sufficient efficiency for replacement of chromates.[368–371] Among synergistic constituents, cathodic inhibitors are synergists of molybdate inhibition.[311] In neutral or alkaline solutions, these cations can interrupt the cathodic reaction of the corrosion process by forming an adherent, insoluble oxide, hydroxide or carbonate film, which is not provided by the oxyanion analogues of chromate. Zn^{2+} most efficiently synergizes molybdate inhibition of steel in aerated, neutral and alkaline cooling tower water.[312] Ca^{2+}, another cathodic inhibitor usually present as hardness in cooling water, significantly increased the corrosion protection of steel already synergistically inhibited with MoO_4^{2-}–Zn^{2+}.[311,313] An amount of 10 percent of calcium or zinc gluconate was found to reduce considerably the amount of molybdate required for the same inhibition effect as observed in molybdate alone.[180] In a comparative study it was found that permanganate increased the corrosion resistance more than molybdate and molybdate more than cerium(III) nitrate for 6061-T6. However, the order was opposite for 2024-T3.[338–339] Many examples of inhibitors that are synergistic with oxyanion analogues of chromate for the protection of ferrous and nonferrous metals are available in the literature.[314–334]

13

Sol-Gels (Ormosils): Properties and Uses

Conversion coatings are applied to metal surfaces to promote both adhesions of organic finishes such as paints and for corrosion protection of the metal substrate. As an alternative to chromate conversion coatings, sol-gel processing grew out of the ceramics field. In this method, soluble metal salts and/or metal organic materials are used to produce a wide variety of mixed metal oxide and metal-oxide-organic composites.[372–375] It is proposed that the only universal processes for treating several Al alloys that are effective in various corrosion environments and are environmentally compliant are coatings consisting of organofunctional and non-organofunctional silanes.[109,179,376–382] These coatings are a promising solution for the corrosion protection of aluminum alloys, which is a key requirement for aircraft as the U.S. Air Force extends the lifetime of its fleet.[383] The downside of epoxy silicate sol-gel coatings when compared to chromate conversion coatings is that the sol-gel films cannot passivate a damaged area.[378]

In 1985, Wilkes *et al.*[384] first reported successful preparation of a new type of organic-inorganic hybrid material by the reaction of tetraethyl ortho-silicate (TEOS) and polydimethyl siloxane (PDMS), which he named "ceramers". At approximately the same time, Schmidt independently reported the successful preparation of new organic-inorganic hybrid materials, which he termed "ormosils" (organically bonded or modified silicates).[385] Ormosils are hybrid organic-inorganic materials formed through the hydrolysis and condensation of organically modified silanes with traditional alkoxide precursors.[386–387] Later on, after other oxides such as ZrO_2 were also bonded to organic groups, Schmidt has also used the term "ormocers".[388]

The sol-gel process, which is mainly based on inorganic polymerization reactions, is a chemical synthesis method initially used for the preparation of inorganic materials, such as glasses and ceramics. Instead of using metal alkoxides as the precursor for the sol-gel reaction, alkoxysilanes are used as the only or one of the precursors, and the organic groups are introduced into the inorganic network through the silicon-carbon bond in an alkoxysilane.[391, 424, 434]

One of the attractive features of the sol-gel process is that it enables the preparation of numerous types of new organic-inorganic materials with improved thermal, mechanical, optical and electrical properties, such as host oxide materials, which are either impossible or extremely difficult to synthesize by any other process.[384–385, 387, 389–391] The numerous applications of these materials include scratch and abrasive-resistant hard coatings and special coatings for polymeric materials, metal and glass surfaces.[391–402] Specifically for mild steel[403] and aluminum 2024 alloys[387,404] widespread uses of these ormosil materials have been reported.

13.1   Types of Sol-Gels

Silanes used for ormosil manufacture are a family of organo-silicon monomers with the general formula $R-Si(OR')_3$, where R is an organofunctional group and R' is usually a methyl or ethyl group. In an aqueous environment, the alkoxy group hydrolyzes to form a silanol R-Si(OH)$_3$, which in turn forms a chemical bond with the hydrated oxide film.[408,415] The other functional group on the silane molecule, R, may bond strongly with the polymer resin base of the paint coating. Introduction of these covalently bonded RSi groups allows chemical modification of the resulting material's properties. The inorganic components tend to impart durability, scratch resistance, and improved adhesion to the metal substrates, while the organic components contribute to increased flexibility, density and functional compatibility with organic polymer paint systems.[390]

Precursors, which generally are di- and trifunctional silanes, span a wide range of sizes, chemical reactivities and functionalities. The use of precursors containing non-hydrolyzable SiC bonds, such as bifunctional or/and trifunctional alkoxysilanes (R'$_n$Si(OR)$_{4-n}$, n = 1 to 3, R = alkyl, R' = organic group), allows introduction of organic groups directly bonded to the polymerlike silica network.[385, 404, 416–419] Trifunctional alkoxysilanes are more commonly used as precursors than other alkoxysilane precursors because a variety of such silanes are commercially available, while bifunctional alkoxysilanes have to be used in the presence of higher functionality precursors in order to form a three-dimensional network.[391,420]

Ormosils can be divided into three categories based on their preparation methods. In type A, the organic, such as a dye, is mixed into the sol-gel liquid solution, such as triethanolamine (TEOA), in alcohol. On gelation, the organic is trapped in the porous silica matrix. It is assumed that no chemical reactions have occurred between the two constituents.[420,421] In type B, a porous oxide gel is first formed in which the porosity and pore size is controlled by heating. An organic solution is then impregnated into the pores of the gel. The organic phase is then solidified via polymerization, and a nanocomposite, such as paramethoxymethamphetamine (PMMA), is formed in silica. Still, no chemical bonds usually exist between the organic and inorganic phases.[420,422] In type C, the organic solution is added to the oxide gel liquid solution, but unlike type A, a chemical bond is formed between two phases or the inorganic oxide precursor may already have a chemically bonded organic group, such as $CH_3Si(OCH_3)_3$ prior to the reaction. Types A, B, and C can further be mixed. The most common system in this class of hybrids is that of polydimethylsiloxan (PDMS) and tetraethoxysilane (TEOS). Together, these various types of ormosils offer a very wide spectrum of chemistry, structures and applications.[423,425]

13.2   Corrosion Inhibition Mechanism of Sol-Gel Coatings

Other than versatile coating formulations and ease of application under normal conditions, ormosil coatings exhibit increased thickness as compared to their inorganic counterparts.[405–406] Thus, sol-gel derived coatings provide good corrosion protection for various metal substrates, such as Fe, Al, and Zn, due to their ability to form a dense barrier to the penetration of water and corrosion initiators to go along with their good adhesion properties and chemical inertness.[407] It should be noted, however, when adsorbed initially, the silane actually is highly hydrophilic. It becomes hydrophobic by loss of water molecules only after the cure of the paint. This hydrophilicity/hydrophobocity dual nature is a unique property of silanes not shown by any other existing interface modifiers. Interfaces modified by silane perform well even under paints that are poor in terms of permeability, porosity or barrier properties, since the hydrophobic nature of organofunctional groups limits the degree of hydration and reduces the degree of adhesion loss.[408] The reduction in adhesion of paints on nonsilane-treated aluminum surfaces after exposure to an aqueous environment is associated with the transformation of the aluminum oxide film beneath the paint coating to a hydrated oxide, which adheres poorly to the aluminum.[409] Thus, the silane processes do not require the same high-cost paint systems as chromates do, which is another advantage of these novel treatments.[410,411]

Another important aspect of corrosion protective coatings is that they should be barriers between the coatings and their environment, but no known coating system stops completely the transport of oxygen, water and corrosive ions to the coatings/metal interface.[412,446–452] Therefore, most corrosion control coating systems are at least two-coat systems, sometimes even three-coat systems, so that the top-coat layer with its hydrophobic polymer composition has the greatest resistance to UV, and the primer and mid-coat adhere to the substrate and each other due to the high crosslink density and wet adhesion properties of the polymers that exist therein. However, the main reason for multiple-layer coating systems overall is the substantial decline in the probability of one defect area overlying another, thus preventing localized corrosion. Therefore, the same final thickness of coating applied by multiple layers will give a significantly better performance than that of a single layer of this thickness.[383]

Despite the fact that sol-gel coatings do not have the self-healing ability of chromate conversion coatings, they still effectively inhibit certain types of corrosion, such as uniform corrosion, provided there is no coating failure, since coating failures may lead to excessive pitting corrosion for aluminum alloys in particular.[182] Corrosion resistance behavior of sol-gels is related to the crosslinking of the polysiloxane to the metal alkoxide with the formation of MOSi bridges and to the formation of polymetallosiloxane-Al interfacial chemical bonds.[402] Thus, it is desirable to improve the chemical interaction between the first monolayer of the coating and the substrate such that electrochemical reactions like the reduction of oxygen are inhibited and bonds may withstand the attack of water and other aggressive species like OH^-.[421,426]

The adsorption of organic compounds on metal substrates is generally achieved by two ways. Organic compounds are either adsorbed from the electrolyte similar to other conventional inhibitors, or adsorbed onto the metal surface by condensation from the vapor phase similar to that of volatile corrosion inhibitors, such as morpholine, hydrazine or hexylamine salts. With no significant electron transfer between the substrate and the adsorbed molecule, this pure electrostatic adsorption process is called physisorption, which is fast and reversible due to low activation energy.[427,428] However, provided that electron transfer occurs due to orbital overlap between a single pair of electrons of the adsorbed molecule and empty bonds of the solid, physisorption becomes chemisorption, which is highly irreversible. Chemisorption is slower than physisorption and it requires higher activation energy. In contrast to physisorption, it is specific for certain metals. On the other hand, the inhibitor should have free single e pairs, or π-electrons for chemisorptions to occur. Based on the Lewis acid-base concept, higher polarizability of the involved heteroatom leads to stronger chemisorption. The inhibitor is then electron donor and the metal is electron acceptor in agreement with the soft and hard acid and base theory (HSAB).[429–431]

Silane coupling is adherence of the organosiloxane-modified natural polymer to the aluminum surface in the form of chemisorption. As a result of this coupling, sol-gel derived thin films highly adhere to metal surfaces, which is confirmed by bond strength measurements in the literature.[109] Chemisorption of silanes is provided by their hydrolysis in humid atmospheres to silanols R_4-nSi(OH)$_n$. Following hydrolysis, condensation occurs through reactions between –OH or –COOH groups on the polymer precursor, the silanol groups from organosiloxane side-chains and hydroxyl species present on the aluminum surface. The hydrolysis of the silane is expected to be the rate-determining step and polymerization begins when hydrolysis is nearly finished.[432] The commonly used silane coupling agents have the structure $X_3Si(CH_2)_nY$, where X represents a group that can hydrolyze, such as methoxy or ethoxy, and Y an organofunctional group such as chlorine, amine, epoxy, or mercaptosubstituted alkyl groups. Nonfunctional silanes are very similar to functional silanes in their structure, except that they have hydrolyzable SiOC bonds on both ends and are better known as crosslinking agents.[110]

13.3   Synthesis of Sol-Gels

Synthesis is typically described by two steps: first hydrolysis of metal alkoxides to produce hydroxyl groups, followed by polycondensation of the hydroxyl groups and residual alkoxyl groups to form a three-dimensional network. These reactions are as follows:[434–442]

Hydrolysis Reactions

$$Si(OR)_4 + H_2O \longleftrightarrow (HO)Si(OR)_3 + ROH \qquad (13.1)$$

$$(OH)Si(OR)_3 + H_2O \longleftrightarrow (HO)_2Si(OR)_2 + ROH \qquad (13.2)$$

$$(HO)_2Si(OR)_2 + H_2O \longleftrightarrow (HO)_3Si(OR) + ROH \qquad (13.3)$$

$$(HO)_3Si(OR) + H_2O \longleftrightarrow Si(OH)_4 + ROH \qquad (13.4)$$

General Hydrolysis Reaction:

$$M(OR)_x + xH_2O \longleftrightarrow M(OH)_x + xROH \qquad (13.5)$$

Alcohol Condensation (Alcoxolation)

$$\equiv Si\text{-}OR + HO\text{-}Si \equiv \longleftrightarrow \equiv Si\text{-}O\text{-}Si \equiv + ROH \qquad (13.6)$$

Water Condensation (Oxolation)

$$\equiv Si\text{-}OH + HO\text{-}Si \equiv \longleftrightarrow \equiv Si\text{-}O\text{-}Si \equiv + HOH \qquad (13.7)$$

General Condensation Reaction:

$$2M(OH)_x \longrightarrow (OH)_{x-1}M\text{-}O\text{-}M(OH)_{x-1} + H_2O \qquad (13.8)$$

The hydrolysis rate is high under an acidic environment relative to that of condensation, and acid catalysts promote the development of more linear or polymer-like molecules in the initial stages. In addition to the pH of the reaction, the concentration of reagents and the size of the alkoxy group can also influence the hydrolysis and condensation reactions through a steric or leaving-group stability effect. As a result, species such as tetramethoxysilane (TMOS) tends to be more reactive than tetraethoxysilane (TEOS).[372,388]

13.3.1   Gelation

The condensation reaction leads to the formation of a sol, which can be cast into films, fibers or blocks and then gelled through continued condensation. The gel phase in sol-gel processing is defined and characterized as a three-dimensional solid "skeleton" enclosing a liquid phase. Both liquid and solid phases are continuous and of colloidal dimensions. The solid phase is typically a condensed polymeric sol where the particles have cross-linked between themselves to form a three-dimensional network.[388]

13.3.2   Drying

When the gelled materials dry, capillary forces cause shrinkage of the flexible skeleton. The skeleton stiffens as it shrinks until the gel can withstand capillary pressures at which point the pores empty, leaving a microporous solid xerogel. Gel films can be formed on a substrate by two methods, immersion and non-immersion (spray, dip, spinon, etc.) Sol-gel based coatings must be designed to contain and deliver soluble non-chrome inhibitors at a rate to maintain effective concentrations in the coating system.[388]

Highly organic films do not adhere to the metal surface well, presumably due to the low inorganic content and insufficient concentrations of SiOH groups to produce covalent SiOAl bonds with the underlying metal surface. In addition, high viscosity ormosils produced using low hydrolysis water content do not flow evenly over the substrate surface, producing differences in texture at regions where gelation occurred. On the other hand, ormosils prepared from high water content do not wet the aluminum surface well due to high surface tension of the mainly aqueous sol, resulting in very thin, unevenly coated films. Therefore, appropriate inorganic/organic ratio and water content are very important for the formation of good quality, corrosion-resistant barrier films highly adherent to the underlying metal substrate.[443]

13.4   Incorporation of Corrosion Inhibitive Pigments into Sol-Gel Coatings

The main protection mechanisms of coatings in general are:[444]

1. Creating a path of extremely high electrical resistance, thus inhibiting anode-cathode reactions.
2. Creating an effective barrier against the corrosion reactants, primarily water and oxygen.
3. Providing an alternative anode for the dissolution process.
4. Passivating the metal surface with soluble pigments.

The first corrosion protection mechanism of organic coatings, that is to create a path of extremely high electrical resistance between anodes and cathodes, is probably the most important one also.[452] This electrical resistance reduces the flow of current available for anode-cathode corrosion reactions. One way to achieve this is to incorporate corrosion protective pigments into the coatings. Inhibitor pigments can increase the electrical resistance in the coating due to their unique physical properties or due to the physical properties of their products they form in the coating.

In addition to the inhibitive pigments, which contain the anodic, cathodic and mixed inhibitor types, those described earlier, there are two more classes of inhibitors commonly incorporated into protective coatings. These two inhibitor pigment types are barrier and sacrificial pigments.

13.4.1   Barrier Pigments

Barrier pigments are chemically inert, flake or plate-like shaped particles, such as micaeous iron oxide (MIO) particles.[453] The term micaceous refers to its particle shape, which is flake-like or lamellar-shaped. In addition to providing a barrier against diffusion of aggressive species through the coating, barrier pigments also provide mechanical reinforcement to the paint film and, when present in the top-coat, they can also block ultraviolet light, thus shielding the binder from this destructive form of radiation.[454–455]

As a result, barrier pigments can be incorporated into primer, intermediate coat, or top-coat since they are chemically inert and do not react with the metal, unlike inhibitive or sacrificial pigments.[454]

13.4.2   Sacrificial Pigments

Sacrificial pigments usually contain zinc in the form of zinc dust in large amounts. When in electrical contact with the steel surface, the zinc film acts as the anode of a large corrosion cell and protects the steel cathode. In other words, zinc sacrificially corrodes instead of steel.[456] In addition to sacrificially corroding, zinc dust also provides barrier action due to formation of its insoluble corrosion products.[457–459]

13.4.3   Inhibitive Pigments

Inhibitive pigments are soluble species, such as molybdates or phosphates, which are carried to the metal surface, where they inhibit corrosion by passivating the substrate surface mostly by forming protective films.[453] Solubility and reactivity are critical parameters for inhibitive pigments. With too much of both parameters, coating degradation occur due to blistering and delamination.

A successful sol-gel coating application for Al 2024-T3 alloy has been developed in a recent study.[414] Enhancement of the corrosion inhibitive properties of this particular sol-gel coating by incorporating inhibitor pigments into its structure remains a challenge, although there have been some recent developments in some recent studies in which this author was associated. In these aforementioned recent studies, only inhibitive pigments were put into test for the purpose of the research, although it is known that inhibitive pigments or their reaction products can act like barrier or sacrificial pigments as well.

14

Corrosion in Engineering Materials

14.1 Introduction

The engineering community is interested in many materials, but the most important and widely used are the structural steel and reinforced concrete. The civil engineer is required to check that these materials withstand not only the tensile and compressive stresses, but also the effects of various complex stress systems and corrosion.

Corrosion of steel is known by engineers as the result of electrochemical reaction when different potentials are developed by electrically connected metal parts in contact with a solution containing free ions. The so-called electrode potential is dependent on the particular metal and the nature of the solution. Comparative values of electrode potentials may be measured against a standard electrode-electrolyte system. For example, if hydrogen is considered of zero V electrode potential, then lead, iron, zinc and aluminum potentials are 0.13, 0.44, 0.75 and 1.66 V, respectively.

When a metal is placed in an electrolyte and different electrode potentials are generated, current flows through the system, causing attack on the more anodic metal (i.e., the metal with the more negative electrode potential). The cathodic metal (i.e., the metal with the more positive electrode potential) remains unattacked. The reaction on the cathodic metal may be deposition of metal, liberation of hydrogen or formation of OH- hydroxyl ions.

Corrosion may also occur without the presence of different electrode potentials if there is an applied electrical current due to the pickup of stray electrical currents from electrical conductors and equipment or the incidence of induced electrical currents.

14.2   Steel Structures

14.2.1   Corrosive Environments

Steel structures may be exposed to a variety of corrosive elements:

1. Water, moisture and humidity
2. Salt-laden air and rain
3. Chemicals from the atmosphere, splashes or spills

14.2.2   The Corrosion Process in Steel Structures

A clear understanding of the corrosion process is essential to understand the steps to inhibit corrosion with protective coatings.

Oxygen combines with iron, the major element in steel, to form rust. This electrochemical process returns the iron metal to the state that it existed in nature-iron oxide. The most common form of iron oxide or iron ore found in nature is hematite (Fe_2O_3), which is equivalent to what we call rust. Iron in iron ore is separated from the oxide to yield usable forms of iron, steel and various other alloys through rigorous electrochemical reduction processes.

The process of combining iron and oxygen, called oxidation, is accompanied by the production of a measurable quantity of electrical current, which is why this is called an electrochemical reaction. For the reaction to proceed, an anode, a cathode and an electrolyte must be present. This is termed a corrosion cell. In a corrosion cell, the anode is the negative electrode where corrosion occurs (oxidation), the cathode is the positive electrode end and the electrolyte is the medium through which an electrical current flows.

14.2.3   Protection Against Corrosion in Steel Structures

Steel members' deterioration may appear either in external corrosion, which would be visible upon inspection, or in undetected changes that reduce its strength.

Recognition of these problems includes factoring a specific amount of tolerance for damage in the design or providing adequate protection system (for example coating and cathodic protection) and/or planned maintenance programs so that such problems do not occur.[469]

Because the interior of a hollow steel section (HSS) is difficult to inspect, some concern has been expressed regarding internal corrosion. In a sealed HSS, internal corrosion cannot progress beyond the point where the oxygen or chemical oxidation is consumed. If fine openings exist at connections, moisture and air can enter the HSS through capillary action or by aspiration due to the partial vacuum that is created if the HSS is cooled rapidly.

Situations where conservative practice would recommend an internal protective coating include: [469]

1. Open HSS where changes in the air volume by ventilation or direct flow of water is possible, and
2. Open HSS subject to a temperature gradient that would cause condensation.

An HSS that is filled or partially filled with concrete should not be sealed. In the event of fire, water in the concrete will vaporize and may create pressure sufficient to burst a sealed HSS. Care should be taken to keep water from remaining in the HSS during or after construction, since the expansion caused by freezing can create pressure that is sufficient to burst an HSS.

14.2.4   Coatings as a Corrosion Control

A coating may be defined as a material which is applied to a surface as a fluid and which forms, by chemical and/or physical processes, a solid continuous film bonded to the surface.

Eliminating any of the reactants in the process can interrupt corrosion. If a barrier is put onto the iron that prevents oxygen and/or water from coming in contact with steel, the corrosion process can be prevented. Steel is not the only surface protected by such barriers. Other alloys and metals such as stainless steel, brass, aluminum and other materials such as concrete, wood, paper, and plastic are also protected from the environment with coatings. Protective coatings that serve as barriers are the principal means of protecting structures.

14.2.5   Corrosion Protection/Surface Protection

Surface protection of steel falls into two categories:

1. corrosion protection; paint, galvanizing, etc. and
2. fire protection.

This discussion is devoted to corrosion protection. The designer should ensure that the corrosion protection system is compatible with other paint and fire protection systems to be used. Several types of paint and methods of application are suitable for shop use. Contact surfaces for nonslip connections or any surfaces to be welded on site must be clearly identified by the designer so that they remain unpainted by the fabricator. Site painting is used for touching up areas damaged during transportation or erection, or to cover site welds or other such details. Whilst the designer may have little influence over the extent of damage, he can reduce the number of site welds, etc., requiring painting. Site painting is time consuming and, therefore, expensive, and it can look unsightly.[470] Paint should be protected during transportation and erection to minimize damage.

The specification of hard, two-pack chemical resistant paint reduces the likely extent of damage, but it is initially more expensive, more difficult to touch up and takes longer to cure. When additional coats of paint are required for decorative purposes, they will generally need to be applied on site, and for convenience damaged paint can be touched up as part of this operation.

Controlling temperature and humidity, and keeping surfaces clean between the applications of coats may prove difficult on site unless the building envelope is sealed before touching up, or the application of additional coats, commences. Site welds should be minimized because they require careful cleaning and degreasing before paint is applied.

14.3   Concrete Structures

The corrosion of metals, especially steel, in concrete has received increasing attention in recent years because of its widespread occurrence in certain types of structures and the high cost of repairs. The corrosion of steel reinforcement is observed in marine structures and chemical manufacturing plants. Also, bridge decks, parking structures and other structures exposed to chlorides have made the problem particularly prominent. The consequent extensive research on factors contributing to steel corrosion has increased our understanding of corrosion, especially concerning the role of chloride ions.

14.3.1.   Corrosion of Reinforcements in Concrete Members

Concrete normally provides reinforcing steel with excellent corrosion protection. The high alkaline environment in concrete results in the formation of a tightly adhering film, which passivates the steel and protects it from corrosion. In addition, concrete can be proportioned to have a low permeability, which minimizes the penetration of corrosion-inducing substances. Low permeability also increases the electrical resistivity of concrete, which impedes the flow of electrochemical corrosion currents. Because of these inherent protective attributes, corrosion of steel does not occur in the majority of concrete elements or structures. Corrosion of steel, however, can occur if the concrete is not of adequate quality, the structure was not properly designed for the service environment, the environment was not as anticipated or changes during the service life of the concrete.[471] The corrosion of steel reinforcement, therefore, is of the greatest concern.

Chloride ions are considered to be the major cause of premature corrosion of steel reinforcement. Chloride ions are common in nature and small amounts are usually unintentionally contained in the mix ingredients of concrete.

Chloride ions also may be intentionally added, most often as a constituent of accelerating admixtures. Dissolved chloride ions also may penetrate unprotected hardened concrete in structures exposed to marine environments or to deicing salts.

14.3.2   Rate of Corrosion

The corrosion rate of steel reinforcement embedded in concrete is strongly influenced by environmental factors. Both oxygen and moisture must be present if electrochemical corrosion is to occur.

Reinforced concrete with significant gradients in chloride ion content is vulnerable to macrocell corrosion, especially if subjected to cycles of wetting and drying.

Other factors that affect the rate and level of corrosion are heterogeneities in the concrete and the steel, pH of the concrete pore water, carbonation of the Portland cement paste, cracks in the concrete, stray currents and galvanic effects due to contact between dissimilar metals.

Design features also play an important role in the corrosion of embedded steel. Mix proportions, depth of cover over the steel, crack control measures and implementation of measures designed specifically for corrosion protection are some of the factors that control the onset and rate of corrosion.

Deterioration of concrete due to corrosion results because the products of corrosion (rust) occupy a greater volume than the steel and exert substantial stresses on the surrounding concrete. The outward manifestations of the rusting include staining, cracking and spalling of the concrete. Concurrently, the cross section of the steel is reduced. With time, structural distress may occur either by loss of bond between the steel and concrete due to cracking and spalling or as a result of the reduced steel cross-sectional area. This latter effect can be of special concern in structures containing high strength pre-stressing steel in which a small amount of metal loss could possibly induce tendon failure.

Corrosion Rate and pH

The corrosion rate of iron is reduced as the pH increases. Since concrete has a pH higher than 12.5, it is usually an excellent medium for protecting steel from corrosion. Only under conditions where salts are present or the concrete cover has carbonated does the steel become vulnerable to corrosion.

14.3.3 Measures to Withstand Corrosion

The research on corrosion to date has not produced a steel or other type of reinforcement that will not corrode when used in concrete and that is both economical and technically feasible. However, research has pointed to the need for the following:

1. quality concrete,
2. careful design,
3. good construction practices,
4. reasonable limits on the amount of chloride in the concrete mix ingredients,
5. use of corrosion inhibitors,
6. use of protective coatings on the steel and
7. use of cathodic protection.

14.3.4   The Importance of Chloride Ions

Concrete can form an efficient corrosion-preventive environment for embedded steel. However, the intrusion of chloride ions in reinforced concrete can cause steel corrosion if oxygen and moisture are also available to sustain the reaction.

Chloride ions may be introduced into concrete in a variety of ways. Some are intentional inclusion as an accelerating admixture, accidental inclusion as contaminants on aggregates or penetration by deicing salts, industrial brines, marine spray, fog or mist.

Incorporation of Chloride Ions in Concrete during Mixing

One of the best known accelerators of the hydration of Portland cement is calcium chloride. Generally, up to 2 percent solid calcium chloride dihydrate based on the weight of cement is added. Chlorides may be contained in other admixtures such as some water-reducing admixtures where small amounts of chloride are sometimes added to offset the set-retarding effect of the water reducer.

In some cases, where potable water is not available, seawater or water with high chloride content is used as the mixing water. In some areas of the world, aggregates exposed to seawater (or that were soaked in seawater at one time) can contain a considerable quantity of chloride salts. Aggregates that are porous can contain larger amounts of chloride.

Chlorides can permeate through sound concrete (i.e., cracks are not necessary for chlorides to enter the concrete).[471,472]

Electrochemical Role of Free Chloride Ions

There are three modern theories to explain the effects of chloride ions on steel corrosion:[471]

(a) The Oxide Film Theory

Some investigators believe that an oxide film on a metal surface is responsible for passivity and, thus, protection against corrosion. This theory postulates that chloride ions penetrate the oxide film on steel through pores or defects in the film easier than do other ions (e.g., SO_4^{2-}). Alternatively, the chloride ions may colloidally disperse the oxide film, thereby making it easier to penetrate.

(b) The Adsorption Theory

Chloride ions are adsorbed on the metal surface in competition with dissolved O_2 or hydroxyl ions. The chloride ion promotes the hydration of the metal ions and, thus, facilitates the dissolution of the metal ions.

(c) The Transitory Complex Theory

According to this theory, chloride ions compete with hydroxyl ions for the ferrous ions produced by corrosion. A soluble complex of iron chloride forms. This complex can diffuse away from the anode, destroying the protective layer of $Fe(OH)_2$ and permitting corrosion to continue. Some distance from the electrode, the complex breaks down, iron hydroxide precipitates and the chloride ion is free to transport more ferrous ions from the anode.[473]

Evidence for this process can be observed when concrete with active corrosion is broken open. A light green semisolid reaction product is often found near the steel, which, on exposure to air, turns black and subsequently rust red in color.

Corrosion process continues, with more iron ions entering into the concrete and reacting with oxygen to form higher oxides that result in a fourfold volume increase. The expansion of iron oxides produces internal stress, which eventually cracks the concrete. Formation of iron chloride complexes may also lead to disruptive forces.

14.3.5   Types of Corrosion Controlling Mechanisms

It is necessary to have both a cathodic and an anodic reaction for a corrosion process to occur. If the cathodic process is the slower process (the one with the larger polarization), the corrosion rate is considered to be cathodically controlled. Conversely, if the anodic process is slower, the corrosion rate is said to be anodically controlled.

In concrete, one or two types of corrosion rate-controlling mechanisms normally dominate. One is cathodic diffusion, where the rate of oxygen diffusion through the concrete determines the rate of corrosion.

The other type of controlling mechanism involves the development of a high resistance path. When steel corrodes in concrete, anodic and cathodic areas may be as much as several feet apart; therefore, the resistance of the concrete may be of great importance.

Cathodic Protection

The principle of cathodic protection is to change the potential of a metal to reduce the current flow and thereby the rate of corrosion. This is accomplished by the application of a protective current at a higher voltage than that of the anodic surface. The current then flows to the original anodic surface, resulting in cathodic reactions occurring there.

The difficulties in using this method, however, are to determine the correct potential to apply to the system and to make sure that it is applied uniformly.

14.3.6   Stray Current Corrosion

Stray electric currents are those that follow paths other than the intended circuit. They can greatly accelerate the corrosion of reinforcing steel. The most common sources of these are electric railways, electroplating plants and cathodic protection systems.

14.3.7 Stress Corrosion Cracking

Stress corrosion is defined as the process in which the damage caused by stress and corrosion acting together greatly exceeds that produced when they act separately.[474]

In stressed steel, a small imperfection caused by corrosion can lead to a serious loss in tensile strength as the corrosion continues at the initial anode area.

Another form of corrosion that is related to stress corrosion cracking is intergranular corrosion. In this case, a gas, usually hydrogen, is absorbed in the iron, causing a loss of ductility and cracking. Other materials that may cause intergranular corrosion are hydrogen sulfide and high concentrations of ammonia and nitrate salts.

The mechanism of how this type of corrosion proceeds is not fully understood; however, it is believed that it involves the reduction in the cohesive strength of the iron.[471]

14.3.8 Effects of the Concrete Environment on Corrosion

Portland Cement

When Portland cement hydrates, the silicates react with water to produce calcium silicate hydrate and calcium hydroxide. The following simplified equations give the main reactions of Portland cement with water.

$$2(3CaO \times SiO_2) + 6H_2O \rightarrow 3CaO \times 2SiO_2 \times 3H_2O + 3Ca(OH)_2$$
$$(14.1)$$

$$2(2CaO \times SiO_2) + 4H_2O \rightarrow 3CaO \times 2SiO_2 \times 3H_2O + Ca(OH)_2$$
$$(14.2)$$

As previously mentioned, the high alkalinity of the chemical environment normally present in concrete protects the embedded steel because of the formation of a protective oxide film on the steel. The integrity and protective quality of this film depends on the alkalinity (pH) of the environment.

Differences in the types of cement are a result of variation in composition, fineness or both, and as such, not all types of cement have the same ability to provide protection of embedded steel. A well hydrated Portland cement may contain from 15 percent to 30 percent calcium hydroxide by weight of the original cement. This is usually sufficient to maintain a solution at a pH about 13 in the concrete independent of moisture content.[475]

The use of blended cements might, under certain circumstances, be detrimental, because of a reduction in alkalinity. However, blended cements can give a substantial reduction in permeability and also an increase in electrical resistivity, especially where a reduction in the water-cement ratio is made possible.

Also, such blended cements may give concrete as much as two to five times higher resistance to chloride penetration than concrete made with Portland cements.

The effects would be beneficial as far as corrosion is concerned and in some circumstances the benefits associated with blended cements more than offset the adverse effects.

Reduction of alkalinity by leaching of soluble alkaline salts with water is an obvious process. Partial neutralization by reaction with carbon dioxide (carbonation), as present either in air or dissolved in water, is another common process.

The silicates are the major components in Portland cement imparting strength to the matrix. No reactions have been detected between chloride ions and silicates. Calcium chloride accelerates the hydration of the silicates when at least 1 percent by weight is added. Calcium chloride seems to act as an accelerator in the hydration of tricalcium silicate, as well as to promote the corrosion of steel.

Also present in Portland cement are C_3A and an aluminoferrite phase reported as C_4AF. The C_3A reacts rapidly in the cement system to cause flash set unless it is retarded. Calcium sulfate is used as the retarder. Calcium sulfate forms a coating of ettringite ($C_3A \times 3CaSO_4 \times 32H_2O$) around the aluminate grains, thereby retarding their reactivity.

Calcium chloride also forms insoluble reaction products with the aluminates in cement. The most commonly noted complex is $C_3A \times CaCl_2 \times xH_2O$, Friedel's salt. The rate of formation of this material is slower than that of ettringite. Chloridealuminate complex forms after ettringite and prevents further reactions of sulfate with the remaining aluminates.[471,476]

Aggregate

The aggregate generally has little effect on the corrosion of steel in concrete. There are exceptions. The most serious problems arise when the aggregates contain chloride salts. This can happen when sand is dredged from the sea or taken from seaside or arid locations. Porous aggregates can absorb considerable quantities of salt.

Care should be exercised when using admixtures containing chloride in combination with lightweight aggregates.

Lightweight aggregates containing sulfides can be damaging to high-strength steel under stress.

Water

A high moisture content will also substantially reduce the rate of diffusion of carbon dioxide and, hence, the rate of carbonation of the concrete. An important effect of the moisture content of concrete is its effect on the electrical resistivity of the concrete. Progressive drying of initially water-saturated concrete results in the electrical resistivity increasing, and steel corrosion would be negligible even in the presence of chloride ions, oxygen and moisture.

14.3.9   Corrosion Inhibiting Admixtures

Numerous chemical admixtures, both organic and inorganic, have been suggested as specific inhibitors of steel corrosion.

Some of the admixtures, however, may retard time of setting of the cement or be detrimental at later ages. Many would be subject to leaching and, hence, less effective in concrete that has lost soluble material by leaching. Among those compounds reported as inorganic inhibitors are potassium dichromates, stannous chloride, zinc and lead chromates, calcium hypophosphite, sodium nitrite and calcium nitrite. Organic inhibitors suggested have included sodium benzoate, ethyl aniline and mercaptobenzothiazole.[471]

With some inhibitors, inhibition occurs only at addition rates sufficiently high enough to counteract the effects of chlorides.

Some of the side effects are low strength, erratic times of setting, efflorescence and enhanced susceptibility to the alkali-aggregate reaction.

14.3.10   Concrete Quality

Concrete will offer more protection against corrosion of embedded steel if it is of a high quality. A low water-cement ratio will slow the diffusion of chlorides, carbon dioxide and oxygen, and, also, the increase in strength of the concrete may extend the time before corrosion induced stresses cause cracking of the concrete. The pore volume and permeability can be reduced by lowering the water cement ratio. The type of cement or use of super plasticizing and mineral admixtures may also be an important factor in controlling the permeability and the ingress of chlorides.

14.3.11   Thickness of Concrete Cover Over Steel

The amount of concrete cover over the steel should be as large as possible, consistent with good structural design, the severity of the service environment and cost.

However, in the case of cement paste, the diffusion of chloride ions into the paste is accompanied by both physical adsorption and chemical binding. These effects reduce the concentration of chloride ion at any particular site and, hence, the tendency for inward diffusion is further reduced.

14.3.12   Carbonation

Carbonation occurs when the concrete reacts with carbon dioxide from the air or water and reduces the pH to about 8.5. At this low pH the steel is no longer passive and corrosion may occur. For high quality concrete, in situations where the rate of carbonation is extremely slow, carbonation is normally not a problem unless cracking of the concrete has occurred or the concrete cover is defective or very thin. Carbonation is not a problem in very dry concrete or in water-saturated concrete. Maximum carbonation rates are observed at about 50 percent water saturation.[471]

14.4   Protection Against Corrosion in Concrete Construction

14.4.1   Introduction

Protection of reinforced concrete structures against steel corrosion requires careful design and construction practice, exclusion of chloride ion from the concrete through surface treatment and direct prevention of steel reinforcement. In the last case, two approaches are possible: to use corrosion-resistant reinforcing steel or to nullify the effects of chloride ions on unprotected reinforcement.

14.4.2   Design and Construction Practices

Through careful design and good construction practices, the protection provided by Portland cement concrete to embedded reinforcing steel can be optimized. It is not the sophistication of the structural design that determines the durability of a concrete member in a corrosive environment but the detailing practices.

The provision of adequate drainage and a method of removing drainage water from the structure are particularly important.

In reinforced concrete members exposed to chlorides and subjected to intermittent wetting, the degree of protection against corrosion is determined primarily by the depth of cover to the reinforcing steel and the permeability of the concrete.

Modern concrete structures should be built with a sufficiently low water-to-binder ratio and a large concrete cover, as these measures do not only increase the time for the chloride to reach the concrete, but also minimize moisture and temperature variations at the steel-concrete interface and, thus, increase the chloride threshold.

The time to spalling is a function of the ratio of cover-to-bar diameter, the reinforcement spacing and the concrete strength. Although conventional Portland cement concrete is not impermeable, concrete with a very low permeability can be made through the use of good quality materials, a minimum water-cement ratio consistent with placing requirements, good consolidation and finishing practices and proper curing.

In concrete that is continuously submerged, the rate of corrosion is controlled by the rate of oxygen diffusion that is not significantly affected by the concrete quality or the thickness of cover. However, corrosion of embedded steel is a rare occurrence in continuously submerged concrete structures.[471,477–479]

Placing limits on the allowable amounts of chloride ion in concrete is an issue still under active debate. Since chlorides are present naturally in most concrete-making materials, specifying zero chloride content for any of the mix ingredients is unrealistic.[471]

However, it is also known that wherever chloride is present in concrete, the risk of corrosion increases as the chloride content increases. When the chloride content exceeds a certain value (termed the "chloride corrosion threshold"), unacceptable corrosion may occur. Provided that other necessary conditions, chiefly the presence of oxygen and moisture, exist to support the corrosion reactions, it is a difficult task to establish a chloride content below which the risk of corrosion is negligible that is appropriate for all mix ingredients and under all exposure conditions and that can be measured by a standard test.

Three different analytical values have been used to designate the chloride content of fresh concrete, hardened concrete or any of the concrete mixture ingredients:[471]

1. Total,
2. Acid-soluble and
3. Water-soluble.

The total chloride content of concrete is measured by the total amount of chlorine. Special analytical methods are necessary to determine it, and acid-soluble chloride is often mistakenly called total chloride. The acid-soluble method is the test method in common use and measures chloride that is soluble in nitric acid. Water-soluble chloride is extractable in water under defined conditions. The result obtained is a function of the analytical test procedure, particularly with respect to particle size, extraction time and temperature, as well as to the age and environmental exposure of the concrete.

It is also important to distinguish clearly between chloride content, sodium chloride content, calcium chloride content or any other chloride salt content. In this report, all references to chloride content pertain to the amount of chloride ion (Cl^-) present. Chloride contents are expressed in terms of the mass of cement unless stated otherwise.

Work at the Federal Highway Administration laboratories showed that for hardened concrete subject to externally applied chlorides, the corrosion threshold was 0.20 percent acid-soluble chlorides. The average content of water-soluble chloride in concrete was found to be 75 percent to 80 percent of the content of acid-soluble chloride in the same concrete.[480]

These investigations show that, under some conditions, a chloride content of as little as 0.15 percent water-soluble chloride (or 0.20 percent acid-soluble chloride) is sufficient to initiate corrosion of embedded steel in concrete exposed to chlorides in service. However, in determining a limit on the chloride content of the mix ingredients, several other factors need to be considered.

The water-soluble chloride content is not a constant proportion of the acid-soluble chloride content. It varies with the amount of chloride in the concrete, the mix ingredients and the test method.

All the materials used in concrete contain some chlorides, and in the case of cement, the chloride content in the hardened concrete varies with cement composition. Although aggregates do not usually contain significant amounts of chloride, there are exceptions. There are reports of aggregates with an acid-soluble chloride content of more than 0.1 percent of which less than one-third is water-soluble when the aggregate is pulverized. Some aggregates, particularly those from arid areas or dredged from the sea, may contribute sufficient chloride to the concrete to initiate corrosion. A limit of 0.06 percent acid-soluble chloride ion in the combined fine and coarse aggregate (by mass of the aggregate) has been suggested with a further proviso that the concrete should not contain more than 0.4 percent chloride (by mass of the cement) derived from the aggregate.

There is thought to be a difference in the chloride corrosion threshold value depending on whether the chloride is present in the mix ingredients or penetrates the hardened concrete from external sources. When chloride is added to the mix, some will chemically combine with the hydrating cement paste.

Conversely, when chloride permeates from the surface of hardened concrete, uniform chloride contents will not exist around the steel, because of differences in the concentration of chlorides on the concrete surface resulting from poor drainage, for example, local differences in permeability, and variations in the depth of cover to the steel.

All these factors promote differences in the environment (oxygen, moisture and chloride content) along a given piece of reinforcement. Furthermore, most structures contain reinforcement at different depths, and, because of the procedures used to fix the steel, the steel is electrically connected. Thus, when chloride penetrates the concrete, some of the steel is in contact with chloride-contaminated concrete while other steel is in chloride-free concrete. This creates a macroscopic corrosion cell that can possess a large driving voltage and a large cathode to small anode ratio which accelerates the rate of corrosion.

ACI 318 allows a maximum water-soluble chloride ion content of 0.06 percent in prestressed concrete, 0.15 percent for reinforced concrete exposed to chloride in service, 1.00 percent for reinforced concrete that will be dry or protected from moisture in service, and 0.30 percent for all other reinforced concrete construction.[481]

The British Code, BS 8110,[482] allows an acid-soluble chloride ion content of 0.35 percent for 95 percent of the test results with no result greater than 0.50 percent. These values are largely based on an examination of several structures in which it was found there was a low risk of corrosion up to 0.4 percent chloride added to the mixture.

The Norwegian Code, NS 3474, allows an acid-soluble chloride content of 0.6 percent for reinforced concrete made with normal Portland cement but only 0.002 percent chloride ion for prestressed concrete.[471]

Corrosion of prestressing steel is generally of greater concern than corrosion of conventional reinforcement because of the possibility that corrosion may cause a local reduction in cross section and failure of the steel. The high stresses in the steel also render it more vulnerable to stress corrosion cracking and, where the loading is cyclic, to corrosion fatigue. However, because of the greater vulnerability and the consequences of corrosion of prestressing steel, chloride limits in the mix ingredients are lower than for conventional concrete.

Normally, concrete materials are tested for chloride content using either the acid-soluble test described in ASTM C 1152, "Acid-Soluble Chloride in Mortar and Concrete," or water-soluble test described in ASTM C 1218, "Water-Soluble Chloride in Mortar and Concrete."

For prestressed and reinforced concrete that will be exposed to chlorides in service, it is advisable to maintain the lowest possible chloride levels in the mix to maximize the service life of the concrete before the critical chloride content is reached and a high risk of corrosion develops. Consequently, chlorides should not be intentionally added to the mix ingredients even if the chloride content in the materials is less than the stated limits. In many exposure conditions, such as highway and parking structures, marine environments and industrial plants where chlorides are present, additional protection against corrosion of embedded steel is necessary.

Since moisture and oxygen are always necessary for electrochemical corrosion, there are some exposure conditions where the chloride levels may exceed the recommended values and corrosion will not occur. Concrete that is continuously submerged in seawater rarely exhibits corrosion induced distress, because there is insufficient oxygen present.

Similarly, where concrete is continuously dry, such as the interior of a building, there is little risk of corrosion from chloride ions present in the hardened concrete. However, interior locations that are wetted occasionally, such as kitchens, laundry rooms or buildings constructed with lightweight aggregate that is subsequently sealed (e.g., with tiles) before the concrete dries out, are susceptible to corrosion damage.

Whereas the designer has little control over the change in use of a building or the service environment, the chloride content of the mix ingredients can be controlled. Estimates or judgments of outdoor "dry" environments can be misleading.

The maximum chloride limits suggested in ACI committee 222 reports are given below. These differ from those contained in the ACI Building Code.

Committee 222 has taken a more conservative approach because of the serious consequences of corrosion, the conflicting data on corrosion threshold values and the difficulty of defining the service environment throughout the life of a structure.

Various nonferrous metals and alloys will corrode in damp or wet concrete. Surface attack of aluminum occurs in the presence of alkali hydroxide solutions. Anodizing provides no protection.

Much more serious corrosion can occur if the concrete contains chloride ions, particularly if there is electrical (metal-to-metal) contact between the aluminum and steel reinforcement, because a galvanic cell is created. Serious cracking or splitting of concrete over aluminum conduits has been reported.

Where concrete will be exposed to chloride, the concrete should be made with the lowest water-cement ratio consistent with achieving maximum consolidation and density. The effects of water-cement ratio and degree of consolidation on the rate of ingress of chloride ions are significant. Concrete with a lower water-cement ratio resists penetration by chloride ions

Table 14.1: Chloride Limit for New Constructions.
(ACI committee 222)[471]

	Acid-soluble, ASTM1152	Water-soluble, ASTM1218
Prestressed concrete	0.08	0.06
Reinforced concrete in wet conditions	0.10	0.08
Reinforced concrete in dry conditions	0.20	0.15

significantly better than concretes with higher water-cement ratios. A low water-cement ratio is not, however, sufficient to insure low permeability. A concrete with a low water cement ratio but with poor consolidation is less resistant to chloride ion penetration than a concrete with a higher water-cement ratio.

ACI 201.2R[477] recommends a minimum of 50 mm (2 in.) cover for bridge decks if the water-cement ratio is 0.40, and 65 mm (2.5 in.) if the water-cement ratio is 0.45. Even greater cover, or the provision of additional corrosion protection treatments, may be required in some environments. These recommendations can also be applied to other reinforced concrete components exposed to chloride ions and intermittent wetting and drying.

Even where the recommended cover is specified, construction practices must insure that the specified cover is achieved. Conversely, placing tolerances, the method of construction and the level of inspection should be considered in determining the specified cover.

The role of cracks in the corrosion of reinforcing steel is controversial. One viewpoint is that cracks reduce the service life of structures by permitting more rapid penetration of carbonation and a means of access of chloride ions, moisture and oxygen to the reinforcing steel. The cracks, thus, accelerate the onset of the corrosion processes and, at the same time, provide space for the deposition of the corrosion products.

The other viewpoint is that while cracks may accelerate the onset of corrosion, such corrosion is localized. Since the chloride ions eventually penetrate even uncracked concrete and initiate more widespread corrosion of the steel, the result is that after a few years' service there is little difference between the amount of corrosion in cracked and uncracked concrete.

Where the crack is perpendicular to the reinforcement, the corroded length of intercepted bars is likely to be no more than three bar diameters. Cracks that follow the line of a reinforcement bar (as might be the case with a plastic shrinkage crack, for example) are much more damaging, because the corroded length of bar is much greater and the resistance of the concrete to spalling is reduced.

For the purposes of design, it is useful to differentiate between controlled and uncontrolled cracks. Controlled cracks are those that can be reasonably predicted from knowledge of section geometry and loading. For cracking perpendicular to the main reinforcement, the necessary conditions for crack control are that there be sufficient steel so it remains in the elastic state under all loading conditions, and that the steel is bonded at the time of cracking (i.e., cracking must occur after the concrete has set).

Examples of uncontrolled cracking are cracks resulting from plastic shrinkage, settlement or an overload condition.

Uncontrolled cracks are frequently wide and usually cause concern, particularly if they are active. However, they cannot be dealt with by conventional design procedures, and measures have to be taken to avoid their occurrence or, if they are unavoidable, to induce them at places where they are unimportant or can be conveniently dealt with, by sealing for example.

14.4.3   Excluding of Chloride Ion from Concrete

Waterproof membranes

Waterproof membranes have been used extensively to minimize the ingress of chloride ions into concrete. Since external sources of chloride ions are waterborne, a barrier to water will also act as a barrier to any dissolved chloride ions.

The requirements for the ideal waterproofing system are straightforward; it should:[471]

1. be easy to install,
2. have good bond to the substrate,

3. be compatible with all the components of the system, including the substrate, prime coat, adhesives, and overlay (where used), and

4. maintain impermeability to chlorides and moisture under service conditions, especially temperature extremes, crack bridging, aging, and superimposed loads.

The number of types of products manufactured to satisfy these requirements makes generalization difficult. Any system of classification is arbitrary, though one of the most useful is the distinction between the preformed sheet systems and the liquid-applied materials. The preformed sheets are manufactured under factory conditions but are often difficult to install, usually require adhesives and are highly vulnerable to the quality of the workmanship at critical locations in the installation. Although it is more difficult to control the quality of the liquid-applied systems, they are easier to apply and tend to be less expensive.

Given the different types and quality of available waterproofing products, the differing degrees of workmanship and the wide variety of applications, it is not surprising that laboratory and field evaluations of membrane performance have also been variable.

Field performance has been found to depend not only on the type of waterproofing material used, but also on the quality of workmanship, weather conditions at the time of installation, design details and the service environment.[471,483–485]

Blistering, which affects both preformed sheets and liquid-applied materials, is the single greatest problem encountered in applying waterproofing membranes.

It is caused by the expansion of entrapped gases, solvents and moisture in the concrete after application of the membrane. The frequency of blisters occurring is controlled by the porosity and moisture content of the concrete and the atmospheric conditions.

Water or water vapor is not a necessary requirement for blister formation, but is often a contributing factor. Blisters may also result from an increase in concrete temperature or a decrease in atmospheric pressure during or shortly after application of membranes.

Membranes can be placed without blisters if the atmospheric conditions are suitable during the curing period.

Once cured, the adhesion of the membrane to the concrete is usually sufficient to resist blister formation. To insure good adhesion, the concrete surface must be carefully prepared and be dry and free from curing membranes, laitance and contaminants, such as oil drippings. Sealing the concrete prior to applying the membrane is possible but rarely practical.

Where the membrane is to be covered (e.g., with insulation or a protective layer), the risk of blister formation can be reduced by minimizing the delay between placement of the membrane and the overlay.

Polymer Impregnation

Polymer impregnation consists of filling some of the voids in hardened concrete with a monomer and polymerizing in situ. Laboratory studies have demonstrated that polymer impregnated concrete (PIC) is strong, durable and almost impermeable. The properties of PIC are largely determined by the polymer loading in the concrete. Maximum polymer loadings are achieved by drying the concrete to remove nearly all the evaporable water, removing air by vacuum techniques, saturation with a monomer under pressure and polymerizing the monomer in the voids of the concrete while simultaneously preventing evaporation of the monomer.

Chemical initiators, which decompose under the action of heat or a chemical promoter, have been used exclusively in field applications. Multifunctional monomers are often used to increase the rate of polymerization.

Since prolonged heating and vacuum saturation are difficult to achieve, and increase processing costs substantially, most field applications have been aimed toward producing only a surface polymer impregnation, usually to a depth of about 25 mm. (1 in.).[486]

There have been a few full-scale applications of PIC to protect reinforcing steel against corrosion, and it must still be considered largely an experimental process. Some of the disadvantages of PIC are that the monomers are expensive and the processing is lengthy and costly. The principal deficiency identified to date has been the tendency of the concrete to crack during heat treatment.[471]

Polymer Concrete Overlays

Polymer concrete overlays consisting of aggregate in a polymer binder have been placed experimentally.

Most monomers have a low tolerance to moisture and low temperatures; hence, the substrate must be dry and in excess of 4°C (40°F). Improper mixing of the two (or more) components of the polymer has been a common source of problems in the field. Aggregates must be dry so as not to inhibit the polymerization reaction.

Workers should wear protective clothing when working with epoxies and some other polymers because of the potential for skin sensitization and dermatitis. Manufacturers' recommendations for safe storage and handling of the chemicals must be followed.[487]

A bond coat of neat polymer is usually applied ahead of the polymer concrete. Blistering, which is a common phenomenon in membranes, has also caused problems in the application of polymer concrete overlays.[488]

Portland Cement Concrete Overlays

Portland cement concrete overlays for new reinforced concrete are applied as part of a two-stage construction process. The overlay may be placed before the first-stage concrete has set or several days later, in which case a bonding layer is used between the two lifts of concrete. The advantage of the first alternative is that the overall time of construction is shortened and costs minimized. In the second alternative, cover to the reinforcing steel can be assured and small construction tolerances achieved, because dead load deflections from the overlay are very small. No matter which sequence of construction is employed, materials can be incorporated in the overlay to provide superior properties, such as resistance to salt penetration and wear and skid resistance, than possible using single-stage construction.

Where the second-stage concrete is placed after the first stage has hardened, sand or water blasting is required to remove laitance and to produce a clean, sound surface.

Resin curing compounds should not be used on the first stage construction because they are difficult to remove.

Several different types of concrete have been used as concrete overlays, including conventional concrete, concrete containing steel fibers and internally sealed concrete.

However, two types of concrete, low-slump and latex-modified, each designed to offer maximum resistance to penetration by chloride ions, have been used most frequently.[471,489]

Low-slump Concrete Overlays

The performance of low-slump concrete is dependent solely on the use of conventional materials and good quality workmanship. The water-cement ratio is reduced to the minimum practical (usually about 0.32) through the use of high cement content (over 470 kg/m^3 or 800 lb/yd^3) and a water content sufficient to produce a slump less than 25 mm (1 in.). The concrete is air-entrained, and a water--reducing admixture or mild retarder is normally used. The use of such a high cement factor and low workability mixture dictates the method of mixing, placing and curing the concrete.

Following preparation of the first-stage concrete, a bonding agent of either mortar or cement paste is brushed into the base concrete just before the application of the overlay. The base concrete is not normally pre-wetted. The overlay concrete is mixed on site, using either a stationary paddle mixer or a mobile continuous mixer, because truck mixers are not suited to producing either the quantity or consistency of concrete required. The concrete must be compacted to the required surface profile using equipment specially designed to handle stiff mixtures. Such machines are much heavier and less flexible than conventional finishing machines and have considerable vibratory capacity. The permeability of the concrete to chloride ions is controlled by its degree of consolidation, which is often checked with a nuclear density meter as concrete placement proceeds.

Wet burlap is placed on the concrete as soon as practicable without damaging the overlay (usually within 20 min. of placing), and the wet curing is continued for at least 72 hr. Curing compounds are not used, since not only is externally available water required for more complete hydration of the cement, but the thin overlay is susceptible to shrinkage cracking and the wet burlap provides a cooling effect by evaporation of the water.

Concrete overlays have been used as a protection against reinforcing steel corrosion in new bridges. In general, the overlays are susceptible to cracking, especially on continuous structures, though this is a characteristic of all rigid overlays.[471]

Latex-modified Concrete Overlays

Latex-modified concrete is conventional Portland cement concrete with the addition of a polymeric latex emulsion. The water of suspension in the emulsion hydrates the cement and the polymer provides supplementary binding properties to produce a concrete with a low water-cement ratio, good durability, good bonding characteristics and a high degree of resistance to penetration by chloride ions, all of which are desirable properties in a concrete overlay.

The latex is a colloidal dispersion of synthetic rubber particles in water. The particles are stabilized to prevent coagulation, and antifoaming agents are added to prevent excessive air entrapment during mixing. Styrenebutadiene latexes have been used most widely. The rate of addition of the latex is approximately 15 percent latex solids by weight of the cement.

The construction procedures for latex-modified concrete (typical thicknesses are 40 and 50 mm.) shall consider the following points:

1. The base concrete must be pre-wetted for at least 1 hr. prior to placing the overlay, because the water aids penetration of the base and delays film formation of the latex.
2. A separate bonding agent is not always used, because sometimes a portion of the concrete itself is brushed over the surface of the base.
3. The mixing equipment must have a means of storing and dispensing the latex.
4. The latex-modified concrete has a high slump so that conventional finishing equipment can be used.
5. Air entrainment of the concrete is believed not required for resistance to freezing and thawing.
6. A combination of moist curing to hydrate the Portland cement and air drying to develop the film forming qualities of the latex are required. Typical curing times are 24 hr. wet curing, followed by 72 hr. of dry curing. The film-formation property of the latex is temperature sensitive and film strengths develop slowly at temperatures below 13°C (55°F). Curing periods at lower temperatures may need to be extended and application at temperatures less than 7°C (45°F) is not recommended.

Hot weather causes rapid drying of the latex-modified concrete, which makes finishing difficult and promotes shrinkage cracking. Placing overlays at night avoid these problems.

Where a texture is applied to the concrete as, for example, grooves to impart good skid resistance, the time of application of the texture is crucial. If applied too soon, the edges of the grooves collapse because the concrete flows.

If the texturing operation is delayed until after the latex film forms, the surface of the overlay tears and, since the film does not reform, cracking often results.

The most serious deficiency reported has been the widespread occurrence of shrinkage cracking in the overlays. Many of these cracks have been found not to extend through the overlay and it is uncertain whether this will impair long-term performance.[471]

14.4.4   Methods of Protecting Reinforcing Steel from Chloride Ions

The susceptibility to corrosion of mild steel reinforcement in common use is not thought to be significantly affected by its composition, grade, or the level of stress.[490] Consequently, to prevent corrosion of the reinforcing steel in a corrosive environment, either the reinforcement must be made of a no corrosive material, or conventional reinforcing steel must be coated to isolate the steel from contact with oxygen, moisture, and chlorides.

Noncorrosive Steels

Natural weathering steels commonly used for structural steelwork do not perform well in concrete containing moisture and chloride and are not suitable for reinforcement. Stainless steel reinforcement has been used in special applications, especially as hardware for attaching panels in precast concrete construction, but is much too expensive to replace mild-steel reinforcement in most applications.

Coatings

Metallic coatings for steel reinforcement fall into two categories: sacrificial and noble or non-sacrificial.

In general, metals with a more negative corrosion potential (less noble) than steel, such as zinc and cadmium, give sacrificial protection to the steel. If the coating is damaged, a galvanic couple is formed in which the coating is the anode. Noble coatings, such as copper and nickel, protect the steel only as long as the coating is unbroken, since any exposed steel is anodic to the coating. Even where steel is not exposed, macrocell corrosion of the coating may occur in concrete through a mechanism similar to the corrosion of uncoated steel.

Nickel, cadmium and zinc have all been shown to be capable of delaying and, in some cases, preventing the corrosion of reinforcing steel in concrete, but only zinc-coated (galvanized) bars are commonly available. Field studies of galvanized bars in service for many years in either a marine environment or exposed to deicing salts have failed to show any deficiencies in the concrete.[491]

Marine studies[492] and accelerated field studies[493] have shown that galvanizing will delay the onset of delamination and spall but will not prevent them.

In general, it appears that only a slight increase in life will be obtained in severe chloride environments. When galvanized bars are used, all bars and hardware in the structure should be coated with zinc to prevent galvanic coupling between coated and uncoated steel.[471]

Fusion-bonded epoxy powder coatings are produced commercially and widely used. The epoxy coating isolates the steel from contact with oxygen, moisture and chloride.

The process of coating the reinforcing steel with the epoxy consists of electrostatically applying finely divided epoxy powder to thoroughly cleaned and heated bars. Many plants operate a continuous production line and many have been constructed specifically for coating reinforcing steel. Integrity of the coating is monitored by a holiday detector installed directly on the production line.

The chief difficulty in using epoxy-coated bars has been preventing damage to the coating in transportation and handling.

Cracking of the coating has also been observed during fabrication where there has been inadequate cleaning of the bar prior to coating or the thickness of the coating has been outside specified tolerances. Padded bundling bands, frequent supports and nonmetallic slings are required to prevent damage during transportation.

Coated tie wires and bar supports are needed to prevent damage during placing. Accelerated time-to-corrosion studies have shown that nicks and cuts in the coating do not cause rapid corrosion of the exposed steel and subsequent distress in the concrete.[471]

Consequently, for long life in severe chloride environments, consideration should be given to coating all the reinforcing steel. If only some of the steel is coated, precautions should be taken to assure that the coated bars are not electrically coupled to large quantities of uncoated steel. A damaged coating can be repaired using a two-component liquid epoxy, but it is more satisfactory to adopt practices that prevent damage to the coating and limit touchup only to bars where the damage exceeds approximately 2 percent of the area of the bar.

A study[494] reported that epoxy-coated reinforcing has less slip resistance than normal mill scale reinforcing; although, for the particular specimens tested, the epoxy-coated bars attained stress levels compatible with ACI development requirements.

14.4.5   Corrosion Control Methods

Chemical Inhibitors

A corrosion inhibitor is an admixture to the concrete used to prevent the corrosion of embedded metal. The mechanism of inhibition is complex, and there is no general theory applicable to all situations.

The compound groups investigated have been primarily chromates, phosphates, hypophosphites, alkalies, nitrites and fluorides. Some of these chemicals have been suggested as being effective; others have produced conflicting results in laboratory screening tests.[471]

Many inhibitors that appear to be chemically effective produce adverse effects on the physical properties of the concrete, such as a significant reduction in compressive strength.

Calcium Nitrite Corrosion Inhibiting Admixture

Corrosion inhibitors protect steel in the presence of chloride ions, varying the rate of the corrosion process by influencing the kinetics and/or thermodynamics of the electrochemical reactions responsible for this process. Inhibitors must retard either the anodic or cathodic reactions or both of them simultaneously.[495]

Calcium nitrite is an anodic inhibitor which is added to the concrete at the batch plant like any other liquid concrete admixture.

The Inhibitor provides protection against both ingressed and admixed chlorides. One of the advantages of calcium nitrite is that its concentration can be determined in both plastic and hardened concrete, which is not the case for other admixtures.

The mechanism by which calcium nitrite prevents corrosion of reinforcing steel can be briefly described as follows: the use of calcium nitrite results in the creation and maintenance of a stronger, flawless and stable passive film on steel embedded in concrete, even in the presence of chloride levels much higher than the critical chloride concentration for corrosion onset for conventional concrete (about 0.9 kg of chloride km/m^3).

Nitrite does not enter into reactions involved in producing the anode, but it reacts with the resulting products of the anode. Thus, it cannot affect the size of the anode. As only monolayers of oxides are involved, essentially no nitrite or hydroxide is consumed in forming the initial protective oxides or hydroxide.

Calcium Nitrite is compatible with other admixture systems, including air entrainment, standard water reducers and super-plasticizers and other products, when each is added separately to the concrete. The nitrite solution is stable at prolonged elevated temperatures and has a freezing point of 15°C. Calcium nitrite has been found to comply with the requirements of BS 6920: Part I: Clause 8: 1990. "Tests of Effect on Water Quality".[495]

Selecting the Addition Rate of Calcium Nitrite

Based upon 13 years of product research prior to market introduction, along with 18 years of field projects and extensive testing history and analysis of chloride data, it has been possible to determine the dosage rate of calcium nitrite required to protect a structure for a given expected chloride level (see Table 14.2). These data have been examined and confirmed by a number of independent authorities.[495]

As a side note underdosage will result in calcium nitrite not providing full protection to the steel. However, numerous studies have shown that underdosage will not promote corrosion.

Overdosage of calcium nitrite will result in more protection and increased service life.

Good quality concrete alone is not enough to provide the service lives required by most of the latest specifications for

Table 14.2 – Calcium Nitrite Dosage.

Calcium Nitrite (L/m³, 30% Solution)	Chloride Ion Protection (Kg/m³, at the rebar level)
10	3.6
15	5.9
20	7.7
25	8.9
30	9.5

reinforced concrete structures built in severe environments. Calcium nitrite delays the initiation time to corrosion and lowers the corrosion rate after the onset of corrosion. Calcium nitrite is a corrosion inhibitor that can provide significant improvement in corrosion resistance when used with good quality concrete. It does not have detrimental effects on the mechanical properties of concrete.

Calcium nitrite should not be viewed as an alternative to the design specification for durable concrete nor as a means to "improve" poor quality concrete. Calcium nitrite corrosion inhibitor in combination with good quality concrete is a viable means to achieving long-term durability. Calcium nitrite can be used combination with other corrosion protection measures.

Cathodic Protection

Steel embedded in concrete is normally passivated due to the highly alkaline (high pH) concrete environment. However, if the potential of the steel is more negative than in any naturally occurring condition, regardless of pH, no steel corrosion occurs (immunity).

The method of providing the highly negative steel potentials required for immunity is referred to as cathodic protection. Cathodic protection (CP) of reinforcing steel has been applied to a large number of concrete structures with corrosion damage for more than 25 years.

Worldwide experience shows that CP prevents further damage in a reliable and economical way for a long time. CP is particularly suitable in cases where chloride contamination has caused reinforcement corrosion.

Although cathodic protection is a viable method of protecting reinforcing steel against corrosion in new construction, most installations to date have been to arrest corrosion in existing structures.

14.5 Corrosion of Unbonded Prestressing Tendons

The causes and effects of corrosion of unbonded single strand tendons are, in several respects, different from those of bonded conventional reinforcing or other post-tensioning systems. Thus, the methods for evaluating and repairing corrosion of single strand tendons are also different in some respects.[496]

For example, since the tendons are largely isolated from the surrounding concrete, they may not be affected by deleterious materials, such as chlorides and moisture in the concrete. However, they also are not passivated by the surrounding concrete and can corrode if water gains access to the inside of the sheathing or anchorage and the grease protection is inadequate.

Measures taken to repair and protect the surrounding concrete may not repair or reduce deterioration of the prestressing steel where corrosion has been initiated. The tendons usually require separate evaluation and repair.

14.5.1 Background

Unbonded tendons in the early systems used bundles of wires or strands, sometimes inaccurately called "cables," of various diameters and protected by grease and paper sheathing that were sometimes applied by hand.

The use of unbonded tendons became more common during the late 1950s and early 1960s as progress was made in establishing design and materials standards. Acceptance of the concept was regional at first and was largely the result of sales efforts and design tutoring by tendon suppliers. The use of post-tensioning increased rapidly during the late 1960s and 1970s as the advantages of the system were demonstrated.

For many types of structures, these advantages included shorter construction time, reduced structural depth, increased stiffness and savings in overall cost. In addition to their use in enclosed buildings, unbonded post-tensioning systems were used in parking structures and slabs on grade, and bonded post-tensioning was used on water tanks, bridges, dams and soil tieback systems. Unbonded multiwire and multistrand tendons have been used extensively in nuclear power structures.

Incidents of corrosion of unbonded single strand tendons began to surface during the 1970s. It had been believed by some that corrosion protection would be provided by the grease during shipping, handling and installation and by the concrete thereafter.

However, the early greases often did not provide the corrosion-inhibiting characteristics that are required in the current Post-Tensioning Institute (PTI) "Specifications for Unbonded Single Strand Tendons."

In the early 1980s, the PTI recognized the structural implications of corrosion and began to implement measures to increase the durability of unbonded posttensioning systems. Relying on experience and practice in the nuclear industry using corrosion inhibiting hydrophobic grease, similar performance standards for grease were incorporated.

In the 1989 edition of ACI 318, "Building Code Requirements for Reinforced Concrete," changes were made to incorporate measures that related the required protection of the tendons and the quality of the concrete to the environmental conditions that could promote corrosion of the post-tensioning.

Structures built prior to the adoption of these new standards, especially those in aggressive environments, are more likely to experience corrosion of the post-tensioning system than those designed and built since.

Tendons that are broken, or are known to be damaged by corrosion, can be repaired or supplemented by any of several methods. The more difficult task is to determine the extent of corrosion damage and the degree to which tendon repairs are needed.

14.5.2   Allowable Tensile Stresses in Concrete

In aggressive environments, designing to minimize cracking was used to improve durability by reducing ingress of corrosive elements. Though a properly greased tendon in an intact sheathing may not be affected at first by a crack in the surrounding concrete, corrosion of nearby conventional reinforcing can cause spalling, which may expose the tendon to physical damage and may then lead to corrosion of the strand.

For consideration of long-term durability and corrosion protection, the maximum allowable tensile stresses in the concrete at service loads, after allowance for all prestress losses, are of most interest.

In the 1977 Code, for an allowable tensile stress up to 12 psi (1.0 MPa), a provision was added requiring that the concrete cover for prestressed and non-prestressed steel be increased for prestressed members exposed to earth, weather or corrosive environments. "Concrete protection for reinforcement" required a 50 percent increase in cover for members exposed to weather, earth, or corrosive environments and with a tensile stress greater than 6 psi (0.5 MPa). The allowable tensile stresses as outlined in the 1977 Code are still effective in later code editions.

14.5.3   Condition Survey

Stains on the surface of the concrete can also provide external evidence of possible internal distress due to corrosion of the post-tensioning system. Grease stains on the soffits of slabs, especially at low points of tendon profiles, can be an indication of unrepaired damage to the tendon sheath as well as shallow concrete cover over the tendons. Such grease staining may be accompanied by water stains or evidence of leaching, indicating water infiltration into the slab tendons.

Visual inspection of exposed end anchorage grout pockets should be performed, especially where exposure to moisture is evident, and correlated with any signs of distress such as those described above. Evidence of shrinkage, cracking, debonding, freeze-thaw damage or rust staining coming from the grout pocket may indicate a potential breach in corrosion protection of the anchorage and post-tensioning tendon.

The most obvious external evidence of corrosion damage is the presence of loops of strand sticking out of the structure. Such loops result when the strand breaks and the elastic energy is released suddenly. The strands typically will erupt from the slab at high points or low points in the tendon profile where concrete cover may be shallow, but occasionally only a single wire will burst through the surface of the concrete. Loops formed by this phenomenon can be anywhere from 1 in. (25 mm) to 2 ft (600 mm) high. Rather than bursting from the structure at some point midway between anchorages, the tendon may also protrude out of the structure a distance of several inches or several feet.

Strand breakage can occur without visible disturbance to the concrete, so the absence of strand loops or projections is not to be taken as an absence of broken tendons.

Most post tensioned structures use higher strength concrete (with higher cracking strength) and/or may incorporate (perhaps unintentionally) a significant degree of restraint or redundancy (i.e., below grade construction or two-way slab construction), so it is possible to have as many as 50 percent of the tendons broken in a beam or in an area of slab without obvious distress.

14.5.4   Repair

In a structure where tendon corrosion has been diagnosed, appropriate means of stopping or slowing the rate of corrosion in the existing tendons should be applied.

Eliminating water intrusion is of primary importance, so concrete repairs should be made and cracks should be sealed. Random cracks can be routed and sealed, but consideration should be given to the application of a waterproofing membrane, possibly incorporating a wearing surface as appropriate, if extensive cracking is present or if there is widespread deficiency of protective concrete cover throughout the structure or a portion of the structure.

14.5.5   Strand Replacement

When a strand has been inadvertently cut or damaged, or when corrosion damage is known or believed to be localized, repairs are often made by replacement of part of the strand between anchorages. The old anchors are reused, and the old wedges are never unlocked. The damaged section of strand is cut away and a new piece of strand spliced onto the ends of the original strand using couplers.

Replacing a strand for its full length and using the original anchors is also possible, but dislodging the old wedges is sometimes difficult and the anchors can be damaged in the process. It is usually advisable to replace the anchors with new ones since this gives the opportunity to improve the system's durability.

Epoxy-coated strand meeting ASTM A822/A may be considered for strand replacement. A smaller diameter strand must be used to accommodate the thickness of the coating (30 to 40 mils, or 0.7 to 1.0 mm). Special anchorages and wedges are required for use with epoxy-coated strand, so existing anchorages have to be replaced.[496]

14.6   Cathodic Protection

In cathodically protecting a structure, a favorable electrochemical circuit is established by installing an external electrode in the electrolyte and passing current (conventional) from that electrode through the electrolyte to the structure to be protected. This current polarizes the potential of the cathodic surfaces (relatively positive) on the steel to that of the anodic (more negative) surfaces.

When this is accomplished, there is no current flow between the formerly anodic and cathodic surfaces and corrosion is arrested. This represents a balanced or equilibrium condition. In normal practice, sufficient current is passed to the surfaces, so that the formerly anodic areas will receive current from the electrolyte and their potential will be shifted to the more negative direction.

There are two ways in which the protective electrochemical circuit can be established. One method uses an electrode made of a metal or alloy more negative than the structure to be protected. For example, if the structure to be protected is constructed of steel, either magnesium, zinc or aluminum may be coupled with the structure. In as much as a protective galvanic cell is set up between the steel and the alloy selected, this method is known as the galvanic anode method of cathodic protection. Also, since the galvanic anodes pass current to the electrolyte, they corrode or sacrifice themselves to protect the structure. Hence, magnesium, zinc and aluminum are termed sacrificial anodes. Sacrificial anodes corrode at relatively high rates. Corrosion rates for magnesium, zinc and aluminum are of the order of 17, 26 and 12 lb. per amp year, respectively.

The high consumption rates, as well as low-driving voltage, are the primary disadvantages of the galvanic anode method of cathodic protection. The open circuit potential between steel and magnesium is on the order of 1 V, while zinc and aluminum are somewhat less. Thus, with this method, it is imperative that a low-resistance circuit be established by installation of many anodes in a low-resistance medium. The anodes installed should also be sized in accordance with their respective consumption rates to provide the necessary design life.

The other way in which the favorable electrochemical circuit can be established is by introducing electrical current from an external source. Because an outside source of current is used, this method is termed impressed current cathodic protection. This method also requires the installation of an external electrode in the electrolyte with the structure to be protected.

However, since the current flow is not dependent on the favorable potential difference between the electrode and the structure to be protected, more noble materials can be selected for the anode. These materials include high-silicon cast iron, graphite and even more noble materials such as platinized titanium or platinized niobium. These metals corrode or are consumed very slowly, less than 1 lb. per amp year.

These anodes are coupled to the structure via the external source of electrical power. This source can be in the form of batteries, thermoelectric generators, generators or photovoltaic cells. Most commonly, however, alternating current line voltage is converted to direct current by a rectifier.

The cathodic protection of reinforced concrete structures is, thus, proven technology, and the problems being currently encountered deal with criteria of protection, design and inspection of the installation.

It should be noted, however, that the reinforcement in many offshore structures is connected to the cathodic protection system used on the exposed steel. This results in protection of the reinforcement and current densities of 0.5 to 1.0 mA/m^2 (0.05 to 0.1 mA/ft^2) have been reported. Thus, cathodic protection of the reinforcement, though unintentional, has been applied in several of the largest offshore structures.

The initial application of cathodic protection to bridge decks was in 1974, and other applications have subsequently been made with encouraging results.[471]

In protecting buried structures or structures exposed in water or in soils, low-resistance electrochemical circuits can normally be established. However, on other structures such as bridge decks, a highly conductive overlay consisting of a coke breeze-asphalt mixture or closely spaced anodes to reduce the circuit resistance and to promote uniform distribution of current to all reinforcement is required.[497,498]

The criteria for protection of steel embedded in concrete are not clearly defined. Most commonly, corrosion engineers use the potential compared to a standard reference cell as the sole criterion. The criterion for steel that is buried or submerged is normally accepted as -0.85 V, or more negative than a copper/CSE (copper-copper sulfate reference electrode).

However, steel embedded in concrete exhibits more noble potentials than exposed steel in the order of 0.2 V to 0.3 V more positive. Therefore, some investigators claim that protection is provided at lower potentials, in the order of –0.5 V with CSE reference.[499]

For steel embedded in concrete exposed to the atmosphere, research has indicated that the –0.85 V criterion may not be attainable. Quite possibly the result may be sufficient current to cause concern about lack of bond.[500]

The possibility of the loss of bond of the reinforcing steel is related to high current densities, at least 25 mA/ft^2.

However, it would be most unusual for a cathodic protection system, typically designed to operate at 2 mA/ft^2 of steel surface, to operate in excess of 25 mA/ft^2 for sufficient time (several years) to cause deterioration in bond strength, unless the potential criterion was applied inappropriately.

Corrosion of steel in concrete is controlled by oxygen access. Polarization of the steel is controlled by cathodic protection.

Concrete is a very alkaline medium and cathodic reaction is the reduction of oxygen to hydroxides. The same is true for the cathodic protection currents. If the current reduces the oxygen faster than it can be replenished, when cathodically protecting steel in concrete, the steel will polarize to a more negative value. If the oxygen supply is great, then to obtain greater degrees of polarization, the current supply must be increased. In some bridge decks, the current required to obtain the criterion of -0.85 V would be such that there would be fear for disbondment even though the half-cell potential was not even close to that value. Thus, cathodic protection of concrete embedded steel is not necessarily a standard procedure.

For concrete that is buried or submerged, probably moisture-saturated, the –0.85 V CSE criterion is easily obtained at current densities as low as 25 m A/ft^2. For bridge decks, where the concrete is comparatively dry and oxygen is abundant, the criteria may be –0.85 V if obtainable with reasonable, current density (probably a maximum of 3 mA/ft^2 of deck surface). If not, a shift of 400 mV for all bridge deck half-cell potentials is a criterion developed from the statistical distribution of half-cell potentials that could change the least negative potential to equal or exceed the most negative half-cell potential.

When using the half-cell potential criterion as developed through the E-log I method, there is a risk that there will be times when the cathodic system will not completely control the corrosion of the steel. For example, if the concrete is near saturation, the steel can usually be polarized with relatively small current densities. Then, if the rectifier is regulated by a half-cell potential and the concrete dries so that oxygen becomes abundant and the polarized potential drifts significantly less negative, it is likely there will be insufficient current capacity to raise the potential to the protective potential value.

Corrosion is caused by the flow of electrons or current. The difference in half-cell potentials is the voltage that causes the current to flow. Once the steel is made cathodic in that it receives current, the current causes oxygen to be reduced. This same amount of current may reduce oxygen faster than it is being replenished and result in polarization with an associated potential change. If the oxygen is replenished at the same rate as it is reduced, no additional polarization will result. Thus, if the amount of current for cathodic protection will make all of the steel cathodic and oxygen reduction is taking place, any greater amount of cathodic protection current will simply be wasted on reducing oxygen.

In addition to disbondment, overprotection can result in hydrogen embrittlement.[471] In acid environments hydrogen ions are reduced at the cathode to atomic hydrogen, which, in turn, combines to form gaseous hydrogen.

When overprotection results, hydrogen gas is formed at a faster rate than the rate of diffusion through the coating, in this instance, concrete. When this occurs, gaseous pressure is developed at the steel-coating interface, which tends to either spall the coating (disbondment) or to diffuse as atomic hydrogen into the metal. When hydrogen diffuses into the metal, it further strains the metal lattice, resulting in reduced ductility and toughness. These phenomena are referred to as hydrogen embrittlement. Normally, hydrogen embrittlement affects high-strength steels only, generally those having yield strengths of 90 ksi (620 MPa) or higher[501,502] and is consequently a potential problem in applying cathodic protection to prestressed concrete elements.

Because of the adverse effects possible from overprotection, polarized potentials (determined immediately after the current has been interrupted) are normally limited to 1.10 V CSE to avoid the possibility of disbonding and hydrogen embrittlement problems. In addition, protection above that level would require more current and a costlier installation without achieving additional protection from corrosion.[471]

Cathodic protection is by far the most versatile method of corrosion control, since it is applicable to any electrically continuous structure within a suitable electrolyte. Inasmuch as the steel embedded in concrete, and not the concrete itself, requires the protection from metallic corrosion, damp concrete serves as a suitable electrolyte, and even structures exposed to the atmosphere, such as bridge decks, can be protected cathodically.

14.6.1   Practical Applications in Tropical Environments and Lessons Gained

Cathodic Protection of Seawater Intake Structures in Petrochemical Plants

Impressed current cathodic protection (ICCP) and sacrificial anode cathodic protection (SACP) systems were designed and installed to control chloride induced corrosion of the steel reinforcement in the atmospherically exposed and submerged parts of the seawater structures, respectively. The design and long-term performance assessment of these systems are described and discussed. The monitoring data collected have suggested that all ICCP systems are performing satisfactorily and meeting the design objectives in controlling the corrosion of the steel reinforcement. The SACP systems generally did not meet the specified criterion of 800 mV Ag/AgC1 current-on steel potentials; however, there have been no signs of corrosion or concrete distress in submerged areas since application of SACP system.[503]

The results showed that a steel current density ranging between 8 and 14 mA/m^2 was sufficient to afford required protection to the steel reinforcement in different structures.

The protection afforded to these structures has residual effect and could last up to 2 months or so when the CP system is turned off. As a result of CP application, the corrosion potentials of the steel reinforcement have been shifted by some 100 to 200 mV in the positive direction.

Design Steel Current Density

According to British, "Code of Practice for Cathodic Protection" (BS7361: Part 1:1991) and National Association of Corrosion Engineers (NACE) recommended practice "Cathodic Protection of Reinforced Steel in Atmospherically Exposed Concrete Structures" (RP0290–90), typical recommended current densities for protection of atmospherically exposed reinforced concrete structures range between 10 and 20 mA/m^2 of steel.[503]

Based on the condition survey results, (i.e., the chloride concentration at the reinforcing steel, extent of corrosion and concrete deterioration) and considering the hot and humid aggressive environment, to which the structures are exposed, each CP system was designed using the following criteria:

a. ICCP System
- Atmospherically exposed sections: 20 mA/m² of steel surface area.
- Buried sections: 20 mA/m² of steel surface area.
- The maximum anode current density used was 110 mA/m² of anode.

b. SACP System
- For steel embedded in concrete: 20 mA/m² of steel surface area.
- For steel exposed to seawater: 60 mA/m² of steel surface area.

System Monitoring

To monitor and assess the performance of CP systems, embeddable Ag/AgCl reference electrodes were embedded into the concrete at representative locations of each structure. About 15 to 20 reference electrodes were installed in each structure (i.e., each electrode covering approximately an area of 150 m² to 200 m²).

For an ICCP system, potential measurements were made at the location of each embedded reference electrode to determine the potential decay after current interruption. Applied current and driving voltage of each independent anode zone were also recorded.

For an SACP system, only current-on potentials were measured and recorded at the location of each reference electrode embedded in submerged areas. There was no provision or facility provided to interrupt the current flowing between the sacrificial anode and the steel reinforcement, to measure instant-off steel potentials.

Performance Assessment of CP Systems

a)   ICCP System

The "100mV Potential Decay" criterion is the most commonly used criterion and recommended practice for performance assessment of CP systems of atmospherically exposed structures. The time allowed for such measurements is usually between four and 24 hours after current interruption. Based on theory and experiment, it is considered that a shift of 100 mV to 150 mV reduces the corrosion rate by an order of magnitude.

The results show that the specified criterion was met at most of the monitoring locations (representing the entire structure) for the seawater structures. This implies that the corrosion rate of the steel reinforcement was significantly reduced throughout the structures and the CP systems are meeting the design objectives.

It is evident from the results that as a result of CP application, free corrosion steel potentials had shifted toward less negative potentials at most of the monitoring locations. At many locations, corrosion potentials had been shifted in the range of –10 mV to –100 mV Ag/AgCl, indicating the restoration of passive conditions on the rebars.

The results indicate that environment around the steel reinforcement has been improved in removing the chlorides and increasing the alkalinity due to application of cathodic protection. This implies and supports the statements that CP is the most appropriate repair method for chloride contaminated structures.

It has been suggested that a design current density of 15 mA/m^2 of steel surface area would be sufficient for such structures. Similarly, the operating voltage of the CP systems was very low and ranged between 1.4 V and 3.5 V.

b)   SACP System

It is evident from current-on potentials that the specified criterion of –800 mV Ag/AgCl was not generally met at most parts of the structures. It appears the SACP systems were not able to deliver sufficient current to achieve the required potentials in those areas. Since drive voltage in such systems is limited and small, this may well be due to incapability of anodes to distribute current adequately to remote areas. No visible damage was detected in submerged areas.

The -800 mV criterion is normally recommended for bare steel structures and pipelines in submerged and below ground conditions. For concrete structures, it is not essential to shift the steel potential to that criterion value. Normally, a negative shift of 300 mV can be considered sufficient to control corrosion of the steel reinforcement.

A final conclusion was that new seawater structures should be constructed with a built-in ICCP system, which would not only be economical but also prevent corrosion of the steel reinforcement from day one, providing a safe design life of more than 50 years.

Cathodic Protection of Structures in Coral Sands in the Presence of Saltwater

Cathodic protection in saline mud and soils to protect steel and aluminum structures such as tanks and piping was used for corrosion control in the Caribbean Islands. The conditions encountered with saline mud and sands are unique to these areas and lend themselves to specialized application of galvanic and impressed current corrosion control.[504]

Corrosion loss to infrastructures in tropical saline environments has been extensive and is an ongoing phenomenon. A unique segment of this activity is corrosion of steel and other metals, such as aluminum, in coral sands and mud.

In locations involving coral sands with salt water intrusion, a unique situation exists for low resistance ground beds that can provide cathodic protection over extensive areas. This applies to pipe, tank bottoms and dock structures.

Properly coated aluminum pipe has performed well in this environment for more than 40 years. This alloy requires active monitoring, but the added benefits of low-cost installation and lack of product contamination provide cost-effective systems.

Coral sands and mud lend themselves to standard cathodic protection materials, resulting in the use of low-cost components for either galvanic or impressed current systems. In such marine environments in semiarid areas, consideration must be given to atmospheric corrosion of components as well as proper monitoring procedures.

The Environment

The conditions prevalent at these locations entail the following:

1. Sea water
2. Brackish water
3. Coral mud
4. Coral sand
5. Combinations of the above

The latter are the usual situations that are encountered and can be under quiescent or turbulent highly oxygenated conditions.

Electrolyte resistivity can vary from clean seawater at 20.5 ohm-cm. to saline mud of 300–500 ohm-cm. to brackish water conditions with resistivity up to 5,000 ohm-cm.

Oxygen content can vary from high level in the splash zone, to areas supporting sulfate-reducing bacteria in the complete absence of oxygen. Furthermore, in many developing countries frequently occurring pollution by sewage is common, which further complicates remedial measures.

A unique condition encountered on land that has been built up from coral deposits is the presence of blowholes, fissures and caves, which augments the penetration of seawater to areas remote from the actual seashore. Knowing that seawater makes for a very extensive, uniform, low resistivity "ground bed" for cathodic protection anodes, the above condition facilitates the design of unique cathodic protection systems.

In the presence of seawater penetration augmented by tidal action, the normally limited drying out of ground beds does not occur. Furthermore, due to the granular nature of coral sands, a gas blockage is not a factor. Consequently, "remote" ground beds of high efficiency can be readily achieved with anodes placed in seawater for protecting structures a considerable distance from such locations.

Typical infrastructures

Typical structures encountered in coral sand and muds are as follows:

a)   Petroleum Product Lines

These can run from several inches in diameter to a 24"/30" diameter and from hundreds of feet up to 29 miles in length. Smaller pipe can be bare, galvanized or poorly coated with coal tar mastic, as well as epoxy or polyethylene tape, which can be field or factory applied.

In many instances, the cathodic protection design entailed the use of shallow "deep wells," approximately 50 feet deep, sunk in coral caves full of seawater.

Alternate ground beds were employed in sea beds adjacent to airport locations. In all cases, very uniform current densities were generally achieved along the entire pipe lengths.

A common cathodic protection design is the use of bracelet anodes that originally were zinc but now, with improved alloys, are usually aluminum. Alternately, high-silicon cast iron anodes mounted on sleds, buried in the sea bed 250 feet from a given pipeline and midway between shore and the spar buoy ship connection have performed well. Anode return cables can be a maintenance problem unless properly secured to the pipelines and buried at least 5 feet into the sea bed, between the pipelines and the anode sled. Anode beds can also be installed in the beach itself, but they must be deep enough to be in the saltwater intrusion area.

High-silicon cast iron anodes must be of the chromium-containing alloy due to chlorine evolution and preferably be of tubular construction. Extreme care must be exercised in protecting cable connections and employing cable jacketing that will stand abrasion. In protected areas, high molecular weight polyethylene will work, but in more aggressive areas, special dual jacketing may be required.

In open seawater, platinum clad niobium or dimensionally stable anodes may be readily employed but must be secured properly and in a sturdy manner. In coral holes or caves, shallow deep wells can employ ¼" rod anodes of the above types. ¼

Foundations for fuel storage tanks vary from the tank bottoms being in oiled sand to being perched on sand foundations as much as 5 feet above grade. For installation on well drained sand pads, many installations do not employ cathodic protection at all.

However, leakage from soil side plate corrosion of tank bottoms has been reported. This situation has occurred when mill scale has been prevalent on the plate.

Two factors have contributed to this condition; firstly, salt laden air has penetrated the underside of the tank due to floor movement of the tank bottom with variations in the amount of product in the tank. Then, the presence of this mill scale creates a small anode/larger cathode condition where the mill scale has cracked at the weld and, with time, allowed penetration at this weld. This, coupled with the reduction in soil resistivity from the chlorides in the air and under tank condensation, augments galvanic corrosion at the break in the mill scale.

At many island facilities, the original cathodic protection system had been carbon anodes, 10–15 feet below grade about 5 feet from the tank edge. These had a tendency to dry out, increasing the ground bed resistance and resulting in loss of protection. As the nature of porous saltwater caves under the tank forms become better understood, the use of shallow deep anode bed design became accepted. Several deep wells (50 feet) could readily protect an entire tank form, excluding any underground piping.

Monitoring of tank bottom steel had been done with copper/copper sulfate reference cells at the tank edge. Maintaining up to –2.0 V at the tank edge usually guaranteed a –0.85 V center potential, with ultimate tank bottom polarization.

b)   Cathodic protection hardware:

1. Rectifiers: Generally employed power supply systems are rectifiers, with occasionally solar power use. The most cost effective rectifiers seeing service are oil immersed units employing 110/220 V A.C. input. Aluminum cases have been used.
2. Cables: Normally, 600 V jacketing of HMWPE insulation is adequate. When abrasion may be a problem, placing cable in PVC conduit is helpful. Cable connections must be "robust" and be adequately insulated.
3. Anodes - Impressed Current: Materials from scrap steel to platinum clad titanium can be employed as anode materials. Since chlorine can be liberated, care must be exercised that gas blockage is avoided. Any accumulation of chlorine will lead to anode-to-cable connection failures. In moving seawater or coral mud, anodes may require support structures. Anode sleds have been used successfully for dock structure and ground beds in the sea floor. This latter application is applied to the protection of pipelines in the seabed.
4. Anodes – Galvanic: Zinc anodes have been used for many years in this environment, in free-flowing seawater as well as buried in silts and sand. These anodes are low in cost and, with well coated structures will provide extended life of up to 20 years. The use of long (4" x 4" x 60") zinc anodes works well for corrugated sheet piling and the internal webs of steel "H" piles. With weld-on rebar-to-anode cores, it provides anodes that can be welded to structures above mean low water. In the last 20 years, new aluminum alloys have been developed that perform at 7.5 lb./amp year,

compared to 23 lb. /amp year for zinc. Galvanic bracelet anodes work well on pipelines and eliminate the possibility of cable failures experienced with impressed current systems. However, good welded connections to the pipes are mandatory or properly clad welded cable must be employed.

Cathodic protection monitoring:

A given cathodic protection system is no better than the monitoring system employed to ensure that the structure is properly polarized and maintained under protective potentials.

For soil conditions, copper/copper sulfate reference cells are used to monitor such systems. In mud or seawater situations, silver/silver chloride cells work best, since copper cells can be contaminated by seawater. Also, zinc reference cells used in seawater and of the proper alloy, will give years of reliable service.

Design parameters:

The less exposed metal surface, the easier it is to protect a given structure. Consequently, a well-coated structure lends itself to fairly rapid polarization at minimum cost. Coatings prone to alkali attack must not be used, since the cathodic surfaces will have an alkaline pH.

Protective potentials are generally –0.85 V to a copper/copper sulfate cell and –0.80 V to a silver/silver chloride cell. Other criteria, such as an electronegative shift of 300 mV or polarization with a potential shift of no more than 100 mV upon current cutoff are also used. National Association of Corrosion Engineers (NACE) standards that cover these criteria are available.

Underground or Submerged Metallic Piping Systems

A lot of work has been done on minimum current densities to achieve and maintain protection. These call for 10–12 mA/sq feet of bare steel area initially to as low as 2 mA/sq feet to maintain the protection after polarization.[504]

Typical corrosion rates in tropical seawater are 4 to 6 mils per year. This value drops off with time, since rust film and calcareous coatings are partially protective.

Variations in corrosion rates are affected by temperature (minimal), oxygen content of the seawater, rate of water movement and the amount of abrasive material in the water.

14.7   Corrosion in Industrial Projects

14.7.1   Corrosion in Oil and Gas Production

Corrosion costs the petroleum industry hundreds of millions of dollars each year. Corrosion considerations and appropriate material selections should be an important part of all action.[505]

Deep hot gas wells, CO_2 floods, deep water offshore platforms and arctic developments are excellent examples of cases that have provided many material and corrosion problems and are expected to continue to do so. The forms of corrosion of most importance in oil and gas production are:

1. Weight loss
2. SCC
3. Corrosion fatigue
4. Galvanic corrosion

Corrosion and materials selections are very important part of every engineering design of equipment used for oil and gas production activities. These become even more important for the severe environments encountered in deep gas, CO_2 floods and offshore and arctic conditions. Many alloys, inhibitors, paints and coating are providing effective in combating corrosion.

Methods that have evolved over the years for combating corrosion include inhibition, material selection, cathodic or anodic protection coating and control of the environment.

Weight – loss Corrosion

Weight-loss corrosion is the most common problem and occurs at some finite rate for all materials. The rate varies greatly, depending on the susceptibility of the material under the environmental conditions to which it is exposed.

In the oil field, water normally present, often as a brine, and the corrosive agents are usually CO_2, H_2S, oxygen, mineral and organic acids or other chemicals that affect the pH or cause oxidizing reaction. Corrosion rates in neutral, low-salinity solutions are normally very low. In contrast, corrosion rates are very high in low-pH solutions that form in presence of acidic materials or high CO_2 partial pressure (Partial pressure is calculated by multiplying mole fraction of CO_2 by the system pressure). Generally the system is considered corrosive if the partial pressure is above 15 psi (103kPa).

H_2S can also cause significant corrosion if the pH is low and oxygen is present. This also usually occurs in the form of pitting or crevice corrosion.

Stress Corrosion Cracking (SCC)

SCC is a dominant factor in the gas fields. The gas form these fields contained high concentrations of H_2S (often greater than 10 percent). Rapid equipment failures occurred in a number of early field tests, which led to a large effort by the petroleum industry and its suppliers to identify and to solve this problem.

Cracking is generally considered more serious than weight loss corrosion because it can happen rapidly and cause loss of fluid containment or render the equipment inoperable.

The cracking forms that are most probable in oil and gas production are SCC and chloride SCC.

H_2S can cause rapid failure of some steels. Generally these steels have yield strengths greater than 100ksi (689 MPa).

The use of alloys must be carefully considered because under some conditions they may be susceptible to other forms of SCC, particularly in the presence of H_2S, saltwater, and sulfur.

Corrosion Fatigue

Corrosion fatigue plays an important role in many production activities, such as sucker-rod and drill-pipe failures. Conditions conducive to corrosion fatigue include the presence of cyclic loads and such corrosives as salt water, oxygen and low pH. A solution to this problem is normally found though control of the environment and reduced loading. Modifying the environment to make it less corrosive may include removing the oxygen, raising the pH or adding corrosion inhibitors. Designing and operating the equipment at lower loads is also beneficial.

Galvanic Corrosion

Galvanic corrosion may be a problem when dissimilar metals are placed in contact with each other and are located in corrosive environment. This problem should be carefully considered during design and material selection. The severity of this problem depends on the environment and the connecting metals. For example, brass connecting to steel may cause rapid corrosion of the steel, whereas aluminum connected to steel may cause rapid corrosion of the aluminum. In contrast, connecting steel to a corrosion-resistant alloy may not cause significant corrosion but should be evaluated.

14.7.2   Corrosion of Carbon Steel and Galvanized Steel in Industrial Projects under Pollutant and Marine Environments

A four year atmospheric corrosion study was conducted for carbon steel and galvanized steel at five different sites around Kuwait.[506] In August 1991, an atmospheric corrosion study was initiated for the state of Kuwait for duration of four years. This study was particularly important for Kuwait and the Gulf region at that time because of the burning oil well fires in Kuwait at the end of the Gulf War. These fires were emitting tons of hydrocarbon substances and pollutant gases in the atmosphere, with the highest concentrations being in Kuwait. These substances and gases are bound to have deleterious effects on existing metallic structures in this part of the world.[506]

Carbon Steel

The main external factors promoting atmospheric corrosion are time of contact of electrolytic film with the metal surface (time of wetness), which depends on climatic conditions (rain, humidity, fog and temperature), and aggressive atmospheric contaminants, such as sulfur dioxide, ammonia, hydrogen sulfide and chlorides.

Rust is the electrochemical corrosion product of iron and steel. In the initial periods of exposure, it forms in a highly dispersed state, containing excess of water, with a mainly amorphous structure and is subjected to dehydration and a crystallization mechanism modifying its density. The composition of rust depends on climate and the nature and concentration of pollutants in the atmosphere.

Values determined for the five test sites showed that rust layers formed are not completely adherent and there is a partial control of corrosion at the metal/oxide interface. Examination showed that the corrosion rate of galvanized steel was much higher in the marine atmospheres than in any type of atmosphere.

Investigators have found that the corrosion rates of both types of materials were greater in a marine environment than in urban or rural atmospheres. The corrosion rate of galvanized steel in the marine atmospheres was found to high at the first months of exposure compared to the other sites.

The concentration of chlorides was expected to be high at the site which located at the sea shore. This high concentration may explain the greater initial degree of corrosion observed on the galvanized steel exposed there, compared to the other exposure sites. The high value of relative humidity, which was expected at these marine sites, combined with the high proportions of chlorides explains the high degree of corrosion found at these sites.

There is a major fossil fuel power plant next to the exposure site which was about 1 kilometer from the seashore. The effect of SO_2 and chlorides on galvanized steel corrosion was found to be detrimental. The surface of the galvanized steel sheets were sparsely covered with white corrosion products that was found to be adherent to the surface.

Effect of Pollutants on Atmospheric Corrosion

Pollutants like SO_2, Cl, H_2S, CO_2, NH_3, NOx and dust, in particular, are the first to greatly enhance the corrosion process. SO_2 is emitted in the atmosphere in large amounts during combustion of all types of sulfur-containing fuels, and concentrations of 5 ppm can be easily attained. The SO_2 is oxidized on moist particles and turned into sulfuric acid. This lends to acidification of the electrolyte layer on the metal surface and consequent stimulation of the corrosion process,

Chlorides occur as particulate matter (e.g., NaCl, $CaCl_2$ or $MgCl_2$) mainly in marine atmosphere. These salts are hygroscopic and promote the electrochemical process of atmospheric corrosion by favoring electrolyte formation at low values of relative humidity. H_2S is extremely reactive and reacts with most technical metals, such as copper, nickel and iron.

Being located in a desert (arid) climate, Kuwait is affected by many dust storms during the year and especially during the summer and spring. The amount of dust in the atmosphere and its composition are variable and depend on many factors. Deposition of dust on metal surfaces is pronounced in the initial stages because of the action of soluble electrolyte-forming components in reducing the critical humidity and providing activating anions that promote the anodic processes.

14.7.3   Dead Sea Structures

A study on corrosion in Dead-Sea-environment structures was conducted.[507] The objectives were:

- To study marine corrosion and its effect on reinforcement steel in concrete due to the severe deterioration of reinforced concrete structure in Dead Sea area.
- To study the effect of water-cement ratio overlay thickness and type of mixing effect on corrosion through the interpretation of potential readings of reinforcement steel bars in concrete.
- The effectiveness of electrochemical chloride extraction (ECE) in rehabilitation of steel reinforcements by applying ECE to concrete structures having severe corrosion.

Electrochemical Chloride Extraction (ECE):

Since steel corrosion is an electrochemical process. Once it occurs in a concrete structure an electrochemical measure, such as electrochemical chloride extraction ECE can stop it or slow it down to a significant extent. The ECE process is as follow:

1. A suitable metal is placed or attached to the surface of a concrete structure,
2. An electrical field is applied between this metal and the embedded steel bars by the passage of a direct current through the concrete in such a manner that the bars become negatively charged and the metal becomes positively charged and
3. The negatively charged chloride ions (Cl^-) in the concrete are drawn away from the steel bars toward the surface of the concrete.

Results and Discussion

Effect of mixing water

The potentials of concrete blocks prepared by Dead Sea water showed more negative potentials than those prepared by tap water. This behavior was observed for specimens with different water-cement ratios and for having different concrete overlay thickness.

Effect of overly thickness

Potential readings interpreted show that steel reinforcements with higher overlay thickness had more positive potentials.

Effect of water-cement ratio

For the same overlay thickness and concrete mix (i.e., type of water used), the 0.7 water-cement ratio gave more negative potential readings compared to both 0.6 and 0.5 water-cement ratios.

Effect of mixing water

Concrete cylinders made by mixing with Dead Sea water showed more negative potential than those mixed with tap water due to:

a) Chloride effect: Higher Cl^- ions concentrations destroy the oxide film on the steel bar that formed on the metal surface. Due to high alkalinity, the Fe^{+2} produced at the steel-concrete interface combines with the OH^- from the cathodic reaction ultimately to produce a stable passive film. Chloride ions in the solution, having the same charge as OH^- ions, complete these anions to combine with the Fe^{+2} cations. The resulting iron chloride complexes are thought to be soluble (unstable); therefore, further metal dissolution is not prevented, and ultimately the buildup of voluminous corrosion products takes place.

b) Corrosion mechanism in samples using tap water as their mix is mainly due to the slow dominating diffusion process.

c) The effect of chlorides is twofold in that both the pH and the electrical resistivity of the concrete are lowered.

Electrochemical Chloride Extraction

a) Electrochemical chloride extraction process began after 170 days of exposure of concrete cylinders to Dead Sea Water, both in order to rehabilitate and protect steel bars against corrosion by removing chloride ions. Potential readings began to increase (i.e., become less negative, more positive) because of the use of a DC-impressed current supply that accelerated the extraction process in four weeks to simulate the conditions of concrete under real, prolonged service life time and reversed the direction of Cl^- ions movement from inner steel bar, leaching it to the outer environment of galvanized steel, which acts as an anode.

b) Effects of the three main variables (mixing water, overlay thickness, and water-cement ratio) are reversed after the application of electrochemical chloride extraction.

Conclusion

1. Increasing the water-cement ratio increases corrosion, holding overlay thickness and type of water used in concrete mix constant. Water-cement ratio samples of 0.5 were last to corrode compared to both 0.6 and 0.7 water-cement ratios.

2. Lower overlay thickness allows higher corrosion of steel bars due to shorter path for Cl^- ions that has to be traveled for the same water-cement ratio and type of water used in the mix.

3. Mixing with Dead Sea water encourages higher corrosion of steel bars at both constant water-cement ratio and overlay thickness.
4. Reinforcement in concrete exposed to Dead Sea water would corrode irrespective of the water-cement ratio used in concrete preparation.
5. The results of ECE showed that electro deposition put the reinforcing bar surface into the repassivation and that it also suppressed the corrosion of the reinforcing bar in concrete due to high alkalinity of the steel bar as a result of applied potential difference.

References

1. *Terminology Relating to Corrosion and Corrosion Testing,* American Society for Testing and Materials Designation G 15–99b (Revised), 2000, 03.02.
2. Fontana, M. G.; *Corrosion Engineering,* New York, McGraw–Hill, 1986.
3. Bardal, E.; *Corrosion and Protection,* Springer, 2003, 2.
4. Davis, J. R.; *Corrosion: Understanding the Basics,* ASM International, 2000, 25.
5. Bardal, E.; *Corrosion and Protection,* Springer, 2003, 6.
6. Fontana, M. G.; Greene, N. D.; *Corrosion Engineering,* New York–Singapore, McGraw–Hill, 1967, 1978, 1986.
7. Bardal, E.; *Corrosion and Protection,* Springer, 2003, 92.
8. Kruger, J.; Long, G. G.; Kuriyama, M.; Goldman, A. J.; *Passivity of Metals and Semiconductors,* Amsterdam, Elsevier Science Publishers, 1983, 163.
9. Strehblow, H. H.; *Mechanisms of Pitting Corrosion in Corrosion Mechanisms in Theory and Practice,* 2nd ed., New York, Basel, Marcel Dekker, 2002.
10. Pou, T. E.; Murphy, O. J.; Bockris, J. O. M.; Tongson, L. L.; Monkowski, M.; *Proceedings, 9th International Congress on Metallic Corrosion,* Toronto, 1984, 2, 141.
11. Bertocci, U.; *Advances in Localized Corrosion: Proceedings of 2nd International Conf. on Localized Corrosion, Orlando, FL, 1987,* National Association of Corrosion Engineers, 1987, 127.
12. Smith, E.; *Dislocations in Solids,* Nabarro, F. R. N. (Ed.); New York, North Holland Publishing Co., 1979, 4, 365.

13. Kaesche, H.; *The Corrosion of Metals; Physico–Chemical Principles and Actual Problems*, Berlin, Springer–Verlag, 1966, 374.

14. Combrade, P.; *Crevice Corrosion of Metallic Materials in Corrosion Mechanisms in Theory and Practice*, 2nd ed., New York, Basel, Marcel Dekker, 2002.

15. Leidheiser, H.; *Coatings in Corrosion Mechanisms*, New York, Basel, Marcel Dekker, 1987, 183–186.

16. Ijsseling, F. P.; *Survey of Literature on Crevice Corrosion*, London, The Institute of Materials, 2000.

17. Wallen, B.; Anderson, T.; "Galvanic Corrosion of Copper Alloys in Contact with a Highly Alloyed Stainless Steel in Seawater," 10th Scandinavian Corrosion Congress, 1986.

18. Dexter, S. C.; "Galvanic Corrosion," University of Delaware Sea Grant College Program, November, 1999.

19. Bardal, E.; Drugli, J. M.; Gartland, P. O.; "A Review: The Behavior of Corrosion Resistant Steels in Seawater," *Corrosion Science*, 1974, 30, 343–353.

20. Valen, S.; Bardal, E.; Rogne, T.; Drugli, J. M.; "New Galvanic Series Based Upon Long Duration Testing in Flowing Seawater", 11th Scandinavian Corrosion Congress, 1989.

21. "Practical Galvanic Series," Army Missile Command Report RSTR–67–11, 1997.

22. *Annual Book of ASTM Standards*, Part 10, Philadelphia, ASTM, 1972.

23. Uhlig, H. H.; *Corrosion and Corrosion Control*, New York, London, John Wiley & Sons, 1971.

24. Bardal, E.; *Corrosion and Protection*, Springer, 2003, 131–132.

25. Hatch, J. E.; *Aluminum: Properties and Physical Metallurgy*, Metals Park, OH ASM International, 1984, 301.

26. Bardal, E.; *Corrosion and Protection*, Springer, 2003, 135–136.

27. Hutchings, I. M.; *The Erosion of Materials by Liquid Flow*, Columbus, Materials Technological Institute of the Chemical Processing Industry, 1986, 25.

28. Bardal, E.; *Corrosion and Protection*, Springer, 2003, 154.

29. Lees, D. J.; *Characteristics of Stress Corrosion Fracture Initiation and Propagation*, Metallurgist and Materials Technology, 1982, 141, 29–38.

30. Bardal, E.; *Corrosion and Protection*, Springer, 2003, 170.

31. Roberge, P. R.; *Handbook of Corrosion Engineering*, New York, McGraw–Hill, 1999.

32. Leygraf, C.; *Atmospheric Corrosion in Corrosion Mechanisms in Theory and Practice*, New York, Hong Kong, Marcel Dekker, 1995, 441.

33. Kucera, V.; Mattson, E.; *Atmospheric Corrosion in Corrosion Mechanisms*, New York, Marcel Dekker, 1987.

34. Henriksen, J. F.; Norwegian Inst. of Air Research Communications, 1985.

35. "Effects of Sulfur Compounds on Materials, Including Historical and Cultural Monuments, in Airborne Sulfur Pollution: Effects and Control," United Nations, New York, 1984.

36. Thiel, P. A.; Madey, T. E.; Surf. Sci. Rep. 1987, 7, 1.

37. Fehlner, F. P.; Graham, M. J.; *Thin Oxide Film Formation on Metals in Corrosion Mechanisms in Theory and Practice*, New York, Hong Kong, Marcel Dekker, 1995, 123.

38. Alwitt, R. S.; *Oxides and Oxide Films*, Diggle, J. W.; Vijh, A. K. (Eds.); New York, Marcel Dekker, 1976, 4, 169.

39. Knotkova, D. et al.; *Atmospheric Corrosion of Metals*, Philadelphia, ASTM STP 767, 1982, 7.

40. Barr, T. L.; *J. Vac. Sci. Technol.*, 1977, 14, 828.

41. Marcus, P.; Hinnen, C.; Olefijord, I.; *Surf. Interface Anal.*, 1993, 2011, 923–929.

42. Nylund, A..; Olefijord, I.; *Surf. Interface Anal.*, 1994, 215, 283–289.

43. Bernard, W. J.; Florio, S. M.; *J. Electrochem. Soc.*, 1985, 132, 231–239.

44. Kobayashi, M.; Niioka, Y.; *Corro. Sci.*, 1990, 31, 237.

45. Robinson, J.; Thomson, G. E.; Shimizu, K.; *Oxide Films on Metals and Alloys*, Herbert, K. R.; Thompson, G. E. (Eds.); Pennington, NJ, Electrochemical Society, 1994, 1.

46. Pourbaix, M.; *Atlas of Electrochemical Equilibria in Aqueous Solutions*, Houston, TX, National Association of Corrosion Engineers, 1974.

47. Mattson, E.; Lindgren, S.; *Metal Corrosion in the Atmosphere*, Philadelphia, American Society for Testing and Materials, 1968, 240.

48. Mattson, E.; *Tek. Tidskr.*, 1968, 98, 767.

49. Strehblow, H.H.; *Korros*, 1976, 27, 792.

50. Kingery, W. D.; Bowen, H. K.; Uhlmann, D. R.; *Introduction to Ceramics*, 2nd Ed. New York, Wiley, 1976, 91 ff.

51. Fehlner, F. P.; Mott, N. F.; *Oxid. Metals*, 1970, 2, 59.

52. Sun, K. H.; *J. Am. Chem. Soc.*, 1947, 30, 277.

53. Nagayama, M.; Cohen, M.; J. *Electrochem. Soc.*, 1962, 109, 781; 1963, 110, 670.

54. Okamoto, G.; *Proceedings of the 5th International Congress of Metal Corrosion*, Houston, National Association of Corrosion Engineers, 1974, 8.

55. Bloom, M. C.; Goldenberg, M.; *Corros. Sci.*, 1965, 5, 623.

56. Foley, C. L.; Kruger, J.; Bechtold, C. J.; J. *Electrochem. Soc.*, 1967, 14, 994.

57. Grimblot, J.; Eldridge, J.M.; J. *Electrochem. Soc.*, 1981, 128, 729.

58. Kubaschewski, O.; Hopkins, B. E.; *Oxidation of Metals and Alloys*, 2nd Ed., London, Butterworths, 1962, 266 ff.

59. Jolly, W. L.; *Modern Inorganic Chemistry*, 2nd Ed., New York, McGraw–Hill, 1991, 238.

60. McBee, C. L.; Kruger, J.; *Electrochem. Acta.*, 1972, 17, 1337.

61. Keller, F.; Edwards, J. D.; *Met. Prog.*, 1948, 54, 195–200.

62. Dignam, M. J.; *The Kinetics of Growth of Oxides, Comprehensive Treatise of Electrochemistry*, New York, Plenum, 1981, 4, 247.

63. Wagner, C.; "Models for Lattice Defects in Oxide Layers on Passivated Iron and Nickel," *Ber. Bunsenges, Phys. Chem.*, 1973, 77, 1090.

64. Blecher, B.; *Aluminum Materials Technology for Automobile Construction*, W. J. Bartz (Ed.); London, Mechanical Engineering Publications, 1993, 43–51.

65. Rogers, T. H.; *Marine Corrosion*, London, George Newnes, 1969.

66. Mor, E.; Bonino, G.; *Proc. 5th Europ. Symp. Corr. Inhibitors*, Ferrara, Italy, 1971, 659–670.

67. *Metals Handbook*, 9th Ed., *Corrosion of Metals*, Metals Park, OH: ASM International, 1987, 13.

68. Iverson, W. P.; "An Overview of the Anaerobic Corrosion of Underground Metallic Structures, Evidence for a New Mechanism," *Underground Corrosion*, ASTM STP 741, American Society for Testing and Materials, 1981.

69. *International 2007 Glossary of Corrosion*, Houston, National Association of Corrosion Engineers, 2007.

70. Dismuke, T.; Coburn, S. K.; Hirsch, C. M.; *Handbook of Corrosion Protection for Steel Pile Structures in Marine Environments*, 1st Ed., Washington, D.C., American Iron and Steel Institute, 1981.

71. Klöppel, H.; Fliedner, A.; Kördel, W.; *Behaviour And ecotoxicology of Aluminium in Soil and Water*, Chemosphere, 1997, 351, 353–363.

72. Romanoff, M.; *Underground Corrosion*, Washington, D.C. National Bureau of Standard, 1957, 579.

73. Wilson, J.; *Petroleum Engineering*, 1983, 6, 25.

74. Craig, D.; *Petroleum Engineering*, 1987, 10, 35.

75. Wu, Y.; *Corrosion*, Houston, National Association of Corrosion Engineers International, 1987, 87, 36.

76. Sharp, S. P.; Yarborough, L.; U.S. Patent: U.S. 4,350,600, 1982.

77. Stanford, J. R.; Campbell, G. D.; U.S. Patent: U.S. 3,959,158, 1976.

78. Sutanto, H.; Smerad, V.A.W.; *SPE Production Engineering*, 1990, 8, 295.

79. Houghton, C. J.; Westmark, R. V.; *Downhole Corrosion Mitigation in Ekofisk (North Sea) field in CO₂ Corrosion in Oil and Gas Production, Selected Papers, Abstracts, and References*; Houston, National Association of Corrosion Engineers Task Group T–1–3, 1984.

80. Duncan, R. N.; "Materials Performance in Khuff Gas Service," *Materials Performance*, 1980, 45–53.

81. Vukasovich, M. S.; Farr, J. P. G.; "Molybdate in Corrosion Inhibition–A Review, "*Polyhedron*, 1986, 551–559.

82. Brown, J. M.; Roberts, H. A.; Herrold, J. J.; "Methods for Inhibiting Corrosion in Cooling Water Systems," U.S. Patent 5,425,914, 1995.

83. Bradbury, D.; Swan, T.; Segal, M. G.; "Application Technique for the Descaling of Surfaces," U.S. Patent 2,877,848, 1988.

84. Schauhoff, S.; Kissel, C.L.; "New Corrosion Inhibitors for High Temperature Applications," *Materials and Corrosion*, 2000, 51, 141–146.

85. Fivizzani, K. P.; "Use of Molybdate as Corrosion Inhibitor in a Zinc/Phosphonate Cooling Water Treatment," U.S. Patent 5,320,779, 1994.

86. Singh, G.; "Inhibition of Mild Steel Corrosion in Acid Mine Waters Containing Ferric Ions," *Br. Corros. J.*, 1988, 23, 4250–4253.

87. Hiller, J. E.; *Korros.*, 1966, 20, 943.

88. Karaivanov, S.; Gawrilov, G.; *Korros.*, 1973, 24, 30.

89. *Clays and Clay Minerals*, 1980, 284, 272–280.

90. Engelhardt, R. et al.; *Neue Huette*, 1971, 16, 593.

91. Misawa, T.; *Corros. Sci.*, 1973, 13, 659.

92. Keller, P.; *Korros.*, 1971, 22, 32.

93. Schwarz, H.; *Korros.*, 1972, 23, 648.

94. Baum, H. et al.; *Neue Huette*, 1974, 19, 423.

95. Keller, P.; *Korros.*, 1969, 20, 102.

96. Misawa, T. et al.; *Corros. Sci.*, 1971, 11, 35.

97. Becker, G. et al.; *Arch. Eisenhuettenwes.*, 1969, 40, 341.

98. Sehmbhi, T. S.; Barnes, C.; Ward, J.J.B.; *Alternatives to Chromate Conversion Coatings for Aluminum*, Wantage, Oxon, BNF Metals Technology Centre.

99. Rozenfeld, L.; *Atmospheric Corrosion of Metals*, Houston, National Association of Corrosion Engineers, 1972.

100. Kaesche, II.; *Localized Corrosion*, Houston, National Association of Corrosion Engineers, 1974, 516.

101. Bird, C. E.; Strauss, F. J.; *Mater. Perform.*, 1976, 1511, 27.

102. Bardal, E.; *Corrosion and Protection*, Springer, 2003, 154.

103. Wernick, S.; Pinner, R.; Sheasbay, P.G.; *The Surface Treatment and Finishing of Aluminum and Its Alloys*, ed., England, ASM International Pub. Ltd., 1987, 1, 269–276.

104. Lomakina, S. V.; Shatova, T. S.; Kazansky, L. P.; *Heteropoly Anions As Corrosion Inhibitors For Aluminum In High Temperature Water*, Moscow, Inst. Of Physical Chemistry, Russian Academy of Sciences, 1993, 1645–1655.

105. *Aluminum Statistical Review for 1999*, Washington, D.C., The Aluminum Association, Inc., 2000.

106. Sanders, R. E.; "Technology Innovation in Aluminum Products," *The Journal of the Minerals*, 2001, 532, 21–25.

107. Staley, J. T.; Liu, J.; Hunt, W. H., Jr.; *Advanced Materials and Processes*, 1997, 152–4, 17–20.

108. Kang, Y.; *Studies on the Corrosion Protection of Metal by Alloying and Organic Inhibitors*, 1997, 51.

109. Kang, Y.; *Studies on the Corrosion Protection of Metal by Alloying and Organic Inhibitors*, 1997, 51.

110. Metroke, T. L.; Parkhill, R. L.; Knobbe, E. T.; "Passivation of Metal Alloys Using Sol–Gel Derived Materials–a Review," *Progress in Organic Coatings*, 2001, 41, 233–238.

111. Van Ooij, W. J.; Song, J.; Subramanian, V.; "Silane–Based Pretreatments of Aluminum and Its Alloys As Chromate Alternatives," *Applied Surface Science*, 2005, 2461, 82–89.

112. Hatch, J. E.; *Aluminum: Properties and Physical Metallurgy*, Metals Park, OH, ASM International, 1984, 354.

113. Hatch, J. E.; *Aluminum: Properties and Physical Metallurgy*, Metals Park, OH, ASM International, 1984, 136.

114. Kending, M. W.; Buchelt, R. G.; "Corrosion Inhibition of Aluminum and Aluminum Alloys by Soluble Chromates, Chromate Coatings, and Chromate Free Coatings," *Corrosion*, 2003, 595, 383.

115. *Corrosion Costs and Preventive Strategies in the United States*, CC Technologies Laboratories, Inc. in Dublin, Ohio with support from the Federal Highway Administration and National Association of Corrosion Engineers, 2001.

116. West, J. M.; *Basic Corrosion and Oxidation*, New York, Halsted Press, 1980.

117. "NACE (National Association of Corrosion Engineers) Glossary of Corrosion Terms," *Mat. Pro.*, 1965, 41, 79.

118. Pictilova, N.; Balezin, S. A.; Baranek, V. P.;, New York, Pergamon Press, 1960.

119. Jones, D.   A.; *Principles and Prevention of Corrosion*, Upper Saddle River, NJ, Prentice–Hall, 1996.

120. Hackermann, N.; *Fundamentals of Inhibitors*, Houston, National Association of Corrosion Engineers Basic Corrosion Course, 1965.

121. Trabanelli, G.; Carassiti, V.; *Advances in Corrosion Science and Technology*, New York, Plenum, 1970, 1, 147.

122. Thomas, J. G.; *Corrosion*, London, Newnes–butterworths, 1976, 2, 183.

123. McCafferty, E.; *Corrosion Control by Coatings*, Princeton, Science Press, 1979, 279.

124. Rosenfeld, L.; *Corrosion Inhibitors*, New York, McGraw–Hill, 1981.

125. Nathan, C. C.; *Corrosion Inhibitors*, Houston, National Association of Corrosion Engineers Publications, 1973.

126. Wilcox, G. D.; Babe, D. R.; Warwick, M. E.; "The Role of Molybdates in Corrosion Prevention," *Corrosion Reviews*, 1986, 63, 336.

127. Batchelor, A. W.; Lam, N. L.; Chandrasekaran, M.; *Materials Degradation and Its Control by Surface Engineering*, 2nd Ed., Imperial College Press, 2002, 119 ff.

128. Lumsden, J.; Szklarska–Smialowska, Z.; *Corrosion*, 1978, 34–35, 169.

129. "Proposed Prohibition Of Hexavalent Chromium Chemicals In Comfort Cooling Towers, Part V, 40 CFR, Part 749", Federal Register, 1988.

130. Roti, J.; Saeder, K.; "A Comprehensive Evaluation of Molybdate Based Cooling Water Treatment Technology, Cooling Tower Institute," Paper TP–88–03, 1988.

131. Stranick, M. A.; Weber, T. R.; "Molybdate Corrosion Inhibition in Waters of Low Oxygen Content," *Amax Report*, L–312–60, 1983.

132.   *AWT Technical Reference and Training Manual*, McLean, VA, Association of Water Technologies Inc., 2001.

133. Koch, G. H.; Brongers, M. P. H.; Thompson, N. G.; Virmani, Y. P.; Payer, J. H.; *Corrosion Costs and Preventative Strategies in the United States*, Houston, National Association of Corrosion Engineers Publications, 2004.

134. Heiyantuduwa, R.; Alexander, M. G.; Mackechnie, J. R.; "Performance of a Penetrating Corrosion Inhibitor in Concrete Affected by Carbonation–Induced Corrosion," *J. Mat. in Civ. Engrg.*, 2006, 186, 842–850.

135. Baiqing, Z.; Xiaowei, W.; Qin, L.; Yisheng, P.; "Performance and Mechanism of a Water Stabilizer for Low Hardness Cooling Water," *Anti–Corrosion Methods and Materials*, 2003, 505, 347–351.

136. Whittemore, M.; LaCosse, G.; Riley, J.; "Composition and Method for Inhibiting Chloride–Induced Corrosion and Limescale Formation on Ferrous Metals and Alloys," U.S. Patents 5,948,267, 1999.

137. Gallant, D.; Simard, S.; "A Study on the Localized Corrosion of Cobalt in Bicarbonate Solutions Containing Halide Ions," *Corrosion Science*, 2005, 47, 1810–1838.

138. Narayanan, T. S. N.; Jegannathan, S.; Ravichandran, K.; "Corrosion Resistance of Phosphate Coatings Obtained by Cathodic Electrochemical Treatment: Role of Anode–Graphite Versus Steel," *Progress in Organic Coatings*, 2006, 554, 355–362.

139. Shimura, Y.; Taya, S.; "Oxygen Scavenger and Boiler Water Treatment Chemical," U.S. Patent 6861032, 2002.

140. Davis, J. R.; *Corrosion: Understanding the Basics*, ASM International, 2000, 182.

141. Ozdemir, L.; *North American Tunneling: 2004, Proceedings of the North American Tunneling Conference, Atlanta, GA*, Leiden, the Netherlands, A.A. Balekma, 2004, 350.

142. Harrop, D.; *Chemical Inhibitors for Corrosion Control*, Clubley, B. G. (Ed.); Cambridge, Royal Society of Chemistry, 1991, 2.

143. Clark, W. J.; McCreery, R. L.; *J. Electrochem. Soc.*, 2002, 149, B379.

144. *ASM Handbook*, 10th edition, Metals Park, OH, ASM Inter., 1994, 5, 405–411.

145. Suzuki, Y.; *Corrosion–Resistant Coatings Technology*, New York, Marcel Dekker, 1989, 5.

146. Zhao, J.; Frankel, G.; McCreery, R. L.; *J. Electrochem. Soc.*, 1998, 45, 2258.

147. Wernick, S.; Pinner, R.; Sheasby, P. G.; *Chemical Conversion Coatings, In the Surface Treatment and Finishing of Aluminum and Its Alloys*, Metals Park, OH, ASM International, 1987, 220.

148. Ooij, W. J.; *Corrosion Protection of Aluminum Alloys by Conversion Systems and Organic Coatings*, Chemtech, 1998, 26–63.
149. Smith, C. J. E.; Baldwin, K. R.; Garette, S.A.; Gibson, M. C.; Hewins, M. A. H.; Lane, P. L.; *Proceedings International Symposium on Aluminum Surfaces Science and Technology*, Antwerp, Belgium, ATB Metallurgie, 1997, XXXVII, 266.
150. Agarwala, V. S.; *Trivalent Chromium Solutions for Applying Chemical Conversion Coatings to Aluminum Alloys*, Naval Air Warfare Center, 1996.
151. "Chemical Conversion Materials for Coating Aluminum and Aluminum Alloys," Military Specification MIL–C–81706.
152. Camprestrini, P.; van Westing, E. P. M.; de Wit, J. H. W.; "Influence of Surface Preparation on Performance of Chromate Conversion Coatings on Alclad 2024 Aluminum Alloy Part I: Nucleation and Growth," *Electrochimica Acta*, 2001, 2553–2571.
153. Osborne, J. H.; Du, J.; Nercissiantz, A.; Taylor, S. R.; Bernard, D.; Bierwagen, G. P.; "Advanced Corrosion–Resistant Aircraft Coatings, prepared under contract no. F33615–96–C–5078," Dayton, OH, Wright–Patterson AFB, Air Force Material Command, 2000.
154. Kending, M. W.; Buchelt, R. G.; "Corrosion Inhibition of Aluminum and Aluminum Alloys by Soluble Chromates, Chromate Coatings, and Chromate–Free Coatings," *Corrosion*, 59, 379–400.
155. Lunn, G.; Sansone, E. B.; *Destruction of Hazardous Chemicals in the Laboratory*, Toronto, John Wiley & Sons, 1990, 63–66.
156. Sax, N. I.; Lewis, R. J. Sr.; *Dangerous Properties of Industrial Materials*, 7th ed., New York, Van Nostrand Reinhold, 1989, 2, 912913.
157. Cohen, S. M.; *Corr.*, 1995, 511, 71–78.
158. "Blue Ribbon Advisory Report, Wright Laboratory," Dayton, OH, Wright Patterson Airforce Base, 1995.
159. Metroke, T. L.; Gandhi, J. S.; Apblett, A.; "Corrosion Resistance of Ormosil Coatings on 2024–T3 Aluminum Alloy," *Progress In Organic Coatings*, 2004, 504, 231–246.
160. O'Brien, P. O.; Kortenkamp, A.; *Transit. Metal Chem.*, 1995, 20, 636.
161. *Chromium and Nickel Welding in Monographs on the Evaluation of Carcinogenic Risks to Humans*, Lyon, France, International Agency for the Research on Cancer, 1990, 49.
162. Suzuki, Y.; *Ind. Health*, 1990, 28, 9.
163. Suzuki, Y.; Fukuda, K.; *Arch. Toxicol.*, 1990, 64, 169.

164. Stearns, D.; Wetterhahn, K.; *Chem. Res. Toxicol.*, 1994, 7, 219.

165. Lay, P.; Levina, A.; *J. Am. Chem. Soc.*, 1998, 120, 6704.

166. U.S. Environmental Protection Agency; "Integrated Risk Information System (IRIS) on Chromium, VI," Washington, DC., National Center for Environmental Assessment, Office of Research and Development, 1999.

167. U.S. Environmental Protection Agency; "Integrated Risk Information System (IRIS) on Chromium, III," Washington, DC., National Center for Environmental Assessment, Office of Research and Development, 1999.

168. U.S. Environmental Protection Agency; "Toxicological Review of Hexavalent Chromium", Washington, DC., National Center for Environmental Assessment, Office of Research and Development, 1998.

169. U.S. Environmental Protection Agency; "Toxicological Review of Trivalent Chromium," Washington, DC., National Center for Environmental Assessment, Office of Research and Development, 1998.

170. Agency for Toxic Substances and Disease Registry (ATSDR); "Toxicological Profile for Chromium," Atlanta, GA, U.S. Public Health Service, U.S. Department of Health and Human Services, 1998.

171. Occupational Safety and Health Administration (OSHA). Occupational Safety and Health Standards, Toxic and Hazardous Substances, "Code of Federal Regulations," 29 CFR 1910.1000., 1998.

172. World Health Organization. Chromium, "Environmental Health Criteria 61" Geneva, Switzerland, 1988.

173. U.S. Department of Health and Human Services. Registry of Toxic Effects of Chemical Substances (RTECS, online database), National Toxicology Information Program, National Library of Medicine, Bethesda, MD, 1993.

174. SAIC; "PM/Toxics Integration: Addressing Co–Control Benefits," Research Triangle Park, NC, Submitted to U.S. Environmental Protection Agency, Office of Air Quality Planning and Standards, 1998.

175. American Conference of Governmental Industrial Hygienists (ACGIH), *1999 TLVs and BEI,. Threshold Limit Values for Chemical Substances and Physical Agents, Biological Exposure Indices,* Cincinnati, OH, 1999

176. National Institute for Occupational Safety and Health (NIOSH), *Pocket Guide to Chemical Hazards, U.S. Department of Health and*

Human Services, Cincinnati, OH, Public Health Service, Centers for Disease Control and Prevention, 1997.

177. Daugherty, M. L.; *Chemical Hazard Evaluation and Communication Group, Biomedical and Environmental Information Analysis Section*, Oak Ridge, TN, Health and Safety Research Division, by Martin Marietta Energy Systems, Inc., for the U.S. Department of Energy under Contract No. DE–AC05–84OR21400, 1992.

178. Hughes, A. E.; Taylor, R. J.; Hinton, B. R.; "Chromate Conversion Coatings on 2024 Al Alloy," *Surf. Interface Anal.*, 1997, 25, 223.

179. Friberg, L.; Nordberg, G. F.; Vouk, V. B.; *Handbook of Toxicology of Metals*, Amsterdam, Elsevier, 1986, 2.

180. Metroke, T. L.; Kachurina, O.; Knobbe, E. T.; "Spectroscopic and Corrosion Resistance Characterization of GLYMOTEOS Ormosil Coatings for Aluminum Alloy Corrosion Inhibition," *Progress in Organic Coatings*, 2002, 44, 295–305.

181. Sarc, O. L.; Kapor, F.; Halle, R.; "Corrosion Inhibition of Carbon Steel in Chloride Solutions by Blends of Calcium Gluconate and Sodium Benzoate," *Materials and Corrosion*, 2000, 51, 147–151.

182. "A New Class of Corrosion Inhibitor," *Specialty Chemicals Magazine*, Bricorr 288, 2001.

183. Metroke, T. L.; Kachurina, O.; Knobbe, E. T.; "Electrochemical and Salt Spray Analysis of Multilayer Ormosil/Conversion Coating Systems for the Corrosion Resistance of 2024–T3 Aluminum Alloys," *Journal of Coatings Technology*, 2002, 74–927, 53–61.

184. Twite, R. L.; Bierwagen, G. P.; *Prog. Org. Coat.*, 1998, 33, 91.

185. Cohen, S. M.; *Corrosion*, 1995, 51, 71.

186. Hinton, B. W. R.; *Met. Finish.*, 1991, 89, 15–55.

187. Kending, M. W.; Buchelt, R. G.; "Corrosion Inhibition of Aluminum and Aluminum Alloys by Soluble Chromates, Chromate Coatings, and Chromate Free Coatings," *Corrosion*, 2003, 595, 379.

188. Cotton, F.A.; Wilkinson, G.; *Advanced Inorganic Chemistry*, 2nd ed., New York, Wiley Interscience, 1966, 818.

189. Frankel, G. S.; *Mechanism of Alloy Corrosion and the Role of Chromate Inhibitors*, First Annual Report, Contract no. F49620–961–0479, Columbus, OH, Ohio State University, 1997.

190. Clark, W. J.; McCreery, R. L.; *J. Electrochem. Soc.*, 2002, 149, B379.

191. Pearlstein, F.; Agarwala, V. S.; *Non–Chromate Conversion Coatings for Aluminum Alloys in Corrosion Protection by Coatings and Surface Modification*, Pennington, NJ, The Electrochemical Society, PV 9328 1993, 199.

192. Katzman, H. A.; Malouf, G. M.; Bauer, R.; Stupian, G. W.; *Appl. Surf. Sci.*, 1979, 23, 416–432.

193. Asami, K.; Oki, Thompson, G.   E.; Wood, G. C.; Ashworth, V.; *Electrochim. Acta*, 1987, 32, 337.

194. Drozda, T.; Maleczki, E.; *Radioanal. Nucl. Chem. Lett.*, 1985, 95, 339.

195. Hagans, P. L.; Haas, C. M.; *Surf. Interf. Anal.*, 1994, 21, 65.

196. Kending, M. W.; Davenport, A. J.; Isaacs, H. S.; *Corros. Sci.*, 1993, 43, 41.

197. Yu, H.; Zhang, G.; Wang, Y.; *Appl. Surf. Sci.*, 1992, 62, 217.

198. Kending, M. W.; Davenport, A. J.; Isaacs, H. S.; *Corros. Sci.*, 1993, 43.

199. Xia, L.; Akiyama, E.; Frankel, G.; McCreery, R.; *J. Electrochem. Soc.*, 2000, 147, 2, 556.

200. Zhao, J.; Frankel, G. S.; McCreery, R. L.; *J. Electrochem. Soc.*, 1998, 145, 2258.

201. Ramsey, J.; McCreery, R. L.; *J. Electrochem. Soc.*, 1999, 146, 4076.

202. Kendig, M.; Addison, R.; Jeanjaquet, S.; *J. Electrochem. Soc.*, 1999, 146, 4419.

203. Sato, N.; *Corrosion*, 1989, 5, 354.

204. Cotton, F.; Wilkinson, G.; *Advanced Inorganic Chemistry*, 5th ed., New York, Wiley, 694.

205. Sinko, J.; U.S. Patents, 5,378,446; 5,176,894.

206. *The Handbook of Chemistry and Physics*, 70th ed., Boca Raton, FL, CRC Press, 1997, 187 ff.

207. Uhlig, H. H.; *Corrosion and Corrosion Control*, 2nd ed, New York, Wiley/Interscience, 1971.

208. Jones, D. A.; *Principles and Prevention of Corrosion*, New York, MacMillan, 1992, 49, 116, 489, 506.

209. Weisberg, H. E.; *Chromate and Molybdate Pigments, Paint and Varnish Production*, 1968.

210. Piens, M.; "Importance of Diffusion in the Electrochemical Action of Oxidizing Pigments," *JCT* 51, 1979, 655.

211. Hagans, P. L.; Haas, C. M.; *Chromate Conversion Coatings, in surface Engineering*, Metals Park, OH, ASM International, 1987, 405.

212. Hughes, E.; Taylor, R. J.; Hinton, B. W. R.; *Surf. Interf. Anal.*, 1997, 25, 405.

213. Ilevbare, G. O.; Scully, J. R.; Yuan, Y.; Kelly, R. G.; "Inhibiting of Pitting Corrosion on Aluminum Alloy 2024–T3: Effect of Soluble Chromate Additions vs. Chromate Conversion Coating," *Corrosion*, 2000, 56, 227.

214. Wernick, S.; Pinner, R.; Sheasby, P. G.; *Chemical Conversion Coatings, in the Surface Treatment and Finishing of Aluminum and Its Alloys*, Metals Park, OH, ASM International, 1987, 220.

215. Blecher, B.; *Aluminum Materials Technology for Automobile Construction*, W. J. Bartz (Ed.); London, Mechanical Engineering Publications, 1993, 43–51.

216. Seo, M.; Sato, N.; *Inhibition in the Context of Passivation, 171, in Reviews on Corrosion Inhibitor Science and Technology*, Houston, National Association of Corrosion Engineers, 1993.

217. Sinko, J.; "Challenges of Chromate Inhibitor Pigments Replacement In Organic Coatings," *Progress In Organic Coatings*, 2001, 42, 273–274.

218. Pourbaix, M.; *Atlas of Electrochemical Equilibria in Aqueous Solutions*, Houston, National Association of Corrosion Engineers, 1974.

219. Bennett, E. O.; Bennett, D. L.; *Tribol. Int.*, 1984, 17, 341.

220. *Consumer Factsheet on: nitrates/nitrites*, Washington, D.C., U.S. Environmental Protection Agency, 2006.

221. Pearlstein, F.; Agarwala, V. S.; "Trivalent Chromium Solutions for Applying Chemical Conversion Coatings to Aluminum Alloys or for Sealing Anodized Aluminum," *Plating and Surface Finishing*, 1994, 50–55.

222. Pearlstein, F.; Agarwala, V. S.; U.S. Patent 5,304,257, 1994.

223. Udy, M. J.; *Chromium: Chemistry of Chromium and Its Compounds*, ACS Monograph Series, New York, Reinhold Publishers, 1956, 177–178.

224. Barnes, C.; Ward, J. J. B.; Sehmbhi, T. S.; Carter, V. E.; *Trans IMF*, 1982, 60, 45.

225. Haaksma, R.; Weir, J. A.; *Proc. The 27th International SAMPE Tech. Conf.*, Albuquerque, 1995, 1074–1082.

226. J. Sinko; *Prog. Org. Coat.*, 2002, 423, 267–282.

227. J. Sinko; *Considerations on the Chemistry and Action Mechanism of Corrosion Inhibitor Pigments in Organic Coatings in 6th Biennial Conf. Organic Coatings*, New Paltz, NY, Institute of Material Science, 2000.

228. Mikhailovski, Y. N.; Berdzenishvili, G. A.; *Prot. Met.*, 1986, 21, 6, 704–711.

229. Kendig, M.; Cunningham, M.; Jeanjaquet, S.; Hardwick, D.; *J. Electrochem. Soc.*, 1997, 11, 3721.

230. Frey, C. U.; Richens, D. T.; Merbach, A. E.; *J. Am. Chem. Soc.*, 1996, 118, 5265.

231. Pepper, S. E.; Bunker, D. J.; Bryan, N. D.; Livens, F. R.; Charnock, J. M.; Pattrick, R. A. D.; Collison, D.; Treatment of radioactive wastes: An Xray Absorption Spectroscopy Study Of The Reaction Of Technetium With Green Rust, *J. of Colloid and Interface Science*, 2003, 268, 2, 408–412.

232. Yamamato, T.; *A Novel Anti–Corrosion Pigment Containing Vanadate/Phosphate*, in *Proc. Advances in Corrosion Protection by Organic Coatings*, PV 8913, Pennington, NJ, The Eletrochem. Soc., 1989, 476.

233. Cook, R. L.; Taylor, S. R.; *Corrosion*, 2000, 56, 321

234. Zein, S.; *J. Appl. Electrochem.*, 2001, 31, 711.

235. Moutarlier, V.; Gigandet, M. P.; Ricq, L.; Pagetti, J.; *Appl. Surf. Sci.*, 2001, 1–2.

236. Mitchell, P. C. H.; Wass, S. A.; Complexes of Molybdenum and Tungsten, *Annu. Rep. Prog. Chem., Sect. A: Inorg. Chem.*, 1991, 88, 127–145.

237. Giordano, N.; Castellam, A.; Bart, J. C. J.; Vaghi, A.; Campadelli, F.; *J. Catal.*, 1975, 37, 204.

238. Zingg, D. S.; Makovsky, L. E.; Tischer, R. E.; Brown, F. R.; Hercules, D. M.; *J. Phys. Chem.*, 1980, 84, 2898.

239. Grunert, W.; Shpiro, E. S.; Feldhaus, R.; Anders, K.; Antoshin, G. V.; Minachev, K. M.; *J. Catl.*, 1987, 107, 522.

240. Chappell, P. J. C.; Kibel, M. H.; Baker, B. G.; *J. Catal.*, 1988, 110, 139.

241. Sax, N. I.; Lewis, R. J., Sr.; *Dangerous Properties of Industrial Materials*, 7th edn., New York, Van Nostrand Reinhold, 1989, 3, 2424.

242. Ashmead, H. J.; *J. Appl. Nutr.*, 1972, 24, 8.

243. Sax, N. I.; *Dangerous Properties of Industrial Materials*, 5th Edition, New York, Reinhold, 1979, 836.

244. Sigel, H.; (Ed.), *Metal Ions in Biological Systems, Carcinogenicity and Metal Ions.*, New York, Marcel Dekker, 1980, 10.

245. Occupational Safety and Health Act, Public Law 91–596, 84 Stat. 1593; 29 U.S.C. 655 et seq., as amended, 1970.

246. Federal Mine Safety Health Act of 1977; 86 U.S.C. 801 et seq., as amended, 1977.

247. ISBN 0–936712–54–6, American Conference of Governmental Industrial Hygienists, 6500 Glenway Avenue, Building D5, Cincinnati, OH 45211.

248. Anderson, J.; *J. Aust. Inst. Agric. Sci.*, 1942, 873.

249. Neenan, M.; *Proc. Soil Sci. Soc.*, Fla, 1953, 13, 178.

250. DeRenzo, E. C.; Kaleita, E.; Heytler, P. G.; Oleson, J. J.; Hutchings, B. L.; Williams, J. H.; *Arch. Biochem. Biophys*, 1953, 45, 247.

251. Richert, D. A.; Westerfield, W. W.; *J. Biol. Chem.*, 1953, 203, 915.

252. Burell, R. J.; Roach, W. A.; Shadwell, A.; *J. Nat. Cancer Inst.*, 1966, 35, 201.

253. Nemenko, B. A.; Moldakulova, M. M.; Borina, S.N.; *Vopr. Onkol.*, 1976, 22, 75.

254. Department of Chemical Etiology and Carcinogenesis, Cancer Inst., Chinese Academy of Medical Sciences, Henan Hydrogeological Team, Lab. Of Henan Geo. Bureau, Health Bureau of Anyang District and Linxian Office of Esophagel Cancer Prevention and Treatment, *Chin. J. Oncol.*, 1980, 2, 29.

255. Luo, X. M.; Lu, S. M.; Liu, Y. Y.; *Chin. J. Epidemiol.*, 1982, 3, 91.

256. Luo, X. M.; *Trace Substances in Environmental Health–XVI*, University of Missouri, 1983, 357.

257. Safe Water Drinking Act, Public Law 93523, 88 Stat. 1660; 42 U.S.C. 300 et seq., as amended, 1974.

258. Resource Conservation and Recovery Act, Public Law 94580, 90 Stat. 95; 42 U.S.C. 3251 et seq., as amended, 1976.

259. Federal Water Pollution Control Act, as amended by the Clean Water Act, Public Law 92500, 86 Stat. 816, 33 U.S.C. 1251 et seq., as amended, 1972.

260. Comprehensive Environmental Response Compensation and Liability Act of (1980), Public Law 96510, 94 Stat. 2767; 42 U.S.C. 9601 et seq., as amended, 1980.

261. Toxic Substances Control Act, Public Law 94469, 90 Stat. 2003; 15 U.S.C. 2601 et seq., as amended, 1976.

262. Opresko, D. M.; Ph.D., Chemical Hazard Evaluation Group, Biomedical and Environmental Information Analysis Section, Health and Safety Research Division, by Martin Marietta Energy Systems, Inc., for the U.S. Department of Energy under Contract No. DE–AC05–84OR21400, Oak Ridge, Tennessee, 1993.

263. "National Association of Corrosion Engineers (NACE) Glossary of Corrosion Terms," *Mater. Perform.*, 1968, 7, 10, 68.

264. Killefer, D. H.; Linz, A.; *Molybdenum Compounds: Their Chemistry and Technology*, New York, Interscience, 1952, 163.

265. Robertson, W. D.; *Chem. Eng.*, 1950, 57, 290.

266. Killefer, D. H.; *Paint, Oil Chem. Rev.*, 1954, 117, 24.

267. Robertson, W. D.; *J. Electrochem. Soc.*, 1951, 98, 94.

268. Choudhury, A. K.; Shome, S. C.; *J. Sci. Ind. Res.*, 1958, 17A, 30.

269. Choudhury, A. K.; Shome, S. C.; *J. Sci. Ind. Res.*, 1959, 18A, 568.

270. Shoen, H. O.; Brand, B. G.; *Off. Dig. Fed. Soc. Paint Technol.*, 1960, 32, 1522.
271. Weisbergy, H. E.; *Paint Varn. Prod.*, 1968, 58, 3, 32.
272. Kronstein, M.; von Burgsdorff, W. A.; Hanan, N.; Weir, D. L.; *Aust. Finish. Rev.*, 1966, 12, 13.
273. Kronstein, M.; U.S. Patent 3,272,663, 1966.
274. Chisolm, S. L.; U.S. Patent 3,311,529, 1967.
275. Kronstein, M.; U.S. Patent 3,528,860, 1970.
276. Schnake, P.; Jensen, D. P.; Albrecht, R. H.; NASA Contract NAS 8–11788, Final Report, 1964.
277. Noda, M.; Nakal, H.; Sasaki, M.; Kanno, Z.; Japanese Patent 79,116,338, 1979.
278. Vukasovich, M. S.; *Lubr. Eng.*, 1980, 36, 708.
279. Vukasovich, M. S.; Robitaille, D. R.; U.S. Patent 4,313,837, 1982.
280. Koh, K. W.; U.S. Patent 4,218,329, 1980.
281. Fette, C. J.; *Lubr. Eng.*, 1979, 35, 625.
282. Vukasovich, M. S.; In *Proceedings of the 3rd Inter. Colloqium on Lubrication in Metal Working, Machining and Metal Forming Processes*, Nellingen, Technishe Akademie Esslingen, 1982.
283. Vukasovich, M. S.; *Lubr. Eng.*, 1984, 40, 456.
284. Bayes, A. L.; US Patent 2,147,395, 1939.
285. Lamprey, H.; US Patent 2,147,409, 1939.
286. Lewis, G.W. Jr.; *Research Project No. 37*, Ann Arbor, MI, Climax Molybdenum Company of Michigan, 1961.
287. Wiggle, R. R.; Hospadaruk, V.; Styloglu, E. A.; *Mater. Perform.*, 1981, 20, 6, 13.
288. Rowe, L.C.;; *Corrosion Inhibitors*, C.C. Nathan (Ed.); Houston, National Association of Corrosion Engineers, 1973, 173.
289. Vukasovich, M. S.; Sullivan, F. J.; *Mater. Perform.*, 1983, 22, 8, 25.
290. Wilson, J. C.; Hirozawa, S. T.; Conville, J. J.; U.S. Patent 4,440,721, 1984.
291. Rao, P. V.; Seetharamaiah, K.; Rama Char, T. L.; *J. Electrochem. Soc. India*, 1974, 23, 1, 7.
292. Wiggle, R. R.; Hospadaruk, V.; Tibaudo, F. M.; Paper No. 810038, *Society of Automotive Engineers*, Detroit, MI, 1981.
293. Vukasovich, M. S.; Sullivan, F. J. (Eds); *Inhibitors and Coolant Corrosivity*, 2nd *International Symposium on Engine Coolants and Their Testing*, Philedelphia, American Society for Testing and Materials, 1984.
294. Hirozawa, S. T.; European Patent Application 0,042,937A1, 1982.
295. Yoshioka, T.; Japanese Patent 8,217,472, 1982.

296. Engelhardt, P. R.; Ventura, E. M.; British Patent Application 8,409,522, 1984.
297. Vukasovich, M. S.; Sullivan, F. J.; *Mater. Perform.*, 1983, 22, 8, 25.
298. Rowe, L. C.; Chance, R. L.; Walker, M. S.; *Mater. Perform.*, 1983, 22, 6, 17.
299. Potter, N. M.; Loranger, R. B.; Vergosen III, H. E.; *American Laboratory*, 1984, 104.
300. Robertson, W. D.; *Chem. Eng.*, 1950, 57, 290.
301. Bregman, J. I.; U.S. Patent 3,024,301, 1962.
302. Hatch, G. B.; *Corrosion*, 1965, 21, 129.
303. Robertson, W. D.; *Chem. Eng.*, 1950, 57, 291.
304. Wilcox, G. D.; Gab, D. R.; Warwick, M. E.; *Corr. Rev.*, 1986, 6, 4, 327–365.
305. Pourbaix, M.; *Atlas of Electrochemical Equilibria*, Oxford, Pergamon Press, 1966, 272.
306. Abdallah, M.; El Etre, A. Y.; Soliman, M. G.; Mabrouk, E. M.; "Some Organic and Inorganic Compounds as Inhibitors for Carbon Steel Corrosion in 3.5 Percent NaCl Solution," *Anti-Corrosion Methods and Materials*, 2006, 53/2, 118–123.
307. Zimin, P. A.; Kazansky, L. P.; Izvest.AN SSSR, *Ser.Khim.*, 1983, 1943.
308. Pryor, M. J.; Cohen, M.; *J. Electrochem. Soc.*, 1953, 100, 203.
309. Stern, M.; *J. Electrochem. Soc.*, 1958, 105, 638.
310. Cartledge, G. H.; *Corrosion*, 1962, 18, 3166.
311. Keddman, M.; Pallotta, C.; *J. Electrochem. Soc.*, 1985, 132, 781.
312. Leidheiser, H.; *J. Corrosion*, 1980, 36, 339.
313. Vukasovich, M. S.; Robitaille, D. R.; *J. Less–Common Met.*, 1977, 54, 437.
314. Robitaille, D. R.; Bilek, J. G.; *Chem. Eng.*, 1976, 83, 12, 79.
315. Mansfeld, F.; Wang, V.; Shih, H.; *J. Electrochem. Soc.*, 1991, 138, 12, 174–175.
316. Hughes, E.; Gorman, J. D.; Paterson, P. J. K.; *Corr. Sci.*, 1996, 38, 11, 1957–1976.
317. Gorman, J. D.; Johnson, S. T.; Johnston, P. N.; Paterson, P. J. K.; Hughes, A. E.; *Corr. Sci.*, 1996, 38, 11, 1977–1990
318. Fedrizzi, L.; Deflorian, F.; Canteri, R.; Fedrizzi, M.; Bonora, P. L.; *Proc. Prog. In the Understanding and Prevention of Corrosion*, Vol. 1, P. Eds., Costay, J.M.; Mercer; A.D. (eds.); Barcelona, 1993, 1, 131–138.
319. Hunn, J. V.; U.S. Patent 3,353,979, 1967.
320. Kirkpatrick, T.; Nilles, J. J.; U.S. Patent 3,677,783, 1972.

321. Moore, F. W.; Robitaille, D. R.; Barry, H. F.; U.S. Patent 3,726,694, 1973.
322. Robitaille, D. R.; Vukasovich, M. S.; Barry, H. F.; U.S. Patent 3,874,883, 1975.
323. Robitaille, D. R.; Vukasovich, M. S.; Barry, H. F.; U.S. Patent 3,969,127, 1976.
324. Vukasovich, M. S.; Sullivan, F. J.; U.S. Patent 4,017,315, 1977.
325. Shibasaki, H.; Tsuchiya, H.; Michikawa, T.; Japanese Patent 7,519,833, 1975.
326. Kerfoot, D. G. E.; U.S. Patent 4,132,667, 1979.
327. Kansai Paint Co. Lts., British Patent 1,415,488, 1975.
328. Nakajo, K.; Kanezewa, K.; Sato, S.; Japanese Patent 7,509,596, 1975.
329. Dai Nippon Tokyo Co. Ltd., British Patent 1,459,069, 1976.
330. Roberts, G. L.; Fessler, Jr. R. G.; U.S. Patent 3,346,604, 1967.
331. Rozenfeld, I. L.; Verdiev, S. Ch.; Kyazimov, A. M.; Bairamov, A. Kh.; *Zashch. Met.* , 1982, 18, 866.
332. Vukasovich, M. S.; *Lubr. Eng.*, 1980, 36, 708.
333. Vukasovich, M. S.; Robitaille, D. R.; U.S. Patent 4,313,837, 1982.
334. Vukasovich, M.S. in *Proceedings of the 3rd Inter. Colloqium on Lubrication in Metal Working, Machining and Metal Forming Processes*, Nellingen, Technishe Akademie Esslingen, 1982.
335. Vukasovich, M. S.; *Lubr. Eng.*, 1984, 40, 456.
336. Buchelt, R. G.; Mmidipally, S. P.; Schmutz, P.; Guan, H.; *Corrosion*, 2002, 58, 3.
337. Bethancourt, M.; Botana, F.; Calvino, J.; Marcos, M.; Rodriguez–Chacon, M.; *Corros. Sci.*, 1998, 11, 1803.
338. Baes, C. F.; Mesmer, R. E.; *Hydrolysis of Cations*, Malabar, FL, Robert E. Kreiger Publishing Co., 1986, 138.
339. Arnott, D. R.; Hinton, B. W. R.; Ryan, N. E.; "Cationic–Film–Forming Inhibitors for the Protection of the AA 7075 Aluminum Alloy Against Corrosion in Aqueous Chloride Solution," *Corrosion*, 1989, 45, 12.
340. Buchheit, R. G.; Drewien, C. A.; Martinez, M. A.; "Chromate–Free Corrosion Resistant Conversion Coatings for Aluminum Alloys," *Advances in Coatings Technologies for Corrosion and Wear Resistant Coatings*, The Minerals, Metals&Materials Society, 1995, 173–182.
341. Mor, E. D.; Bonino, G.; *Proceed. 3rd Eur. Symp. Corros. Inhibitors*, Ferrara, Italy, 1970, 659.
342. Kadek, V. M.; Lepin, L. K.; *Proc. 3rd Europ. Symp. Corr. Inhibitors*, Ferrara, Italy, 1970, 643.
343. Lahodny–Sarc, O.; *Proc. 8th Europ. Symp. Corr. Inhibitors*, Ferrara, Italy, 1995, 421.

344. Wruble, C.; Mor, E. D.; Montini, U.; *Proc. 6th Europ. Symp. Corr. Inhibitors*, Ferrara, Italy, 1985, 557.
345. Lahodny–Sarc, O.; Orlovic–Leke, P.; *Proc. 11th Internat. Corr. Congress*, Florence, Italy, 1990, 3, 17.
346. Roti, J. S.; Thomas, P. A.; *Proceed. Internat. Corros. Forum*, New Orleans, 1984, 318.
347. Lahodny–Sarc, O.; Popov, S.; *Surface and Coatings Technology*, 1998, 34, 537.
348. Krasts, E.; Kadek, V.; Klavina, S.; *Proc. 7th Europ. Symp. Corr. Inhibitors*, Ferrara, Italy, 1990, 569.
349. Lahodny–Sarc, O.; Orlovic–Leke, P.; *RAD Crotian Acad. Sc.&Arts*, 1991, 9, 11.
350. Krasts, H. B.; Kadek, V. M.; Lepin, L. K.; *Proc. 4th Europ. Symp. Corr. Inhibitors*, Ferrara, Italy, 1975, 204.
351. Mor, E.D.; Wruble, C.; *Br. Corrosion J.*, 1976, 11, 199.
352. Lahodny–Sarc, O.; *Proc. 5th Europ. Symp. Corr. Inhibitors*, Ferrara, Italy, 1980, 609.
353. Lahodny–Sarc, O.; *RAD Yug. Acad. Sc. &Arts*, 1982, 394, 18.
354. Kadek, V. M.; Krasts, H. B.; *Corrosion Inhibitors on the Basis of Polyhydroxy–complexes of Boric Acid*, 2273–R 1000 JT 05129, 1983.
355. Lahodny–Sarc, O.; Orlovic–Leke, P.; *Proceed. 9th Eur. Congress on Corrosion*, Utrecht, The Netherlands, FU133, 1989.
356. Li, H. M.; *Proceed. Ibid.*, 1990, 3, 3147.
357. Lahodny–Sarc, O.; Orlovic–Leke, P.; *Proceed. 7th Eur. Symp. Corros. Inhibitors*, Ferrara, Italy, 1990, 1025.
358. Lahodny–Sarc, O.; Orlovic–Leke, P.; *Proceed. U.K. Corrosion and Eurocorr.'94*, Bournemouth, U.K., 1994, 1, 120.
359. Mor, E.D.; Wrubl, C.; *Br. Corrosion J.*, 1983, 18, 142.
360. Lahodny–Sarc, O.; Orlovic–Leke, P.; Skanst, V.; *Proceed. U.K. Corrosion and Eurocorr.'88*, Brighton, U.K., 1988, 1, 97.
361. Lahodny–Sarc, O.; Popov, S.; *Surface and Coatings Technology*, 1988, 34, 537.
362. Rajendran, S.; Apparao, B. V.; Palaniswamy, N.; *Br. Corrosion J.*, 1998, 33, 315.
363. Rajendran, S.; Apparao, B. V.; Palaniswamy, N.; *Symposium on Corrosion Control by Coatings, Cathodic Protection and Inhibitors in Seawater*, Dubrovnik, Crotia, 23rd Event of the European Federation of Corrosion, 1998.
364. Talalina, A. S.; Kochanova, L. G.; Ananeva, A. I.; Romanova; A. A.; *Corrosion Inhibitors in the Sterilization and Disinfecting of Medical Instruments*, Meditsinskaya Tekhnika, 1984, 4, 30–33.
365. Opresko, D. M.; Chemical Hazard Evaluation Group, Biomedical and Environmental Information Analysis Section, Health and

Safety Research Division, by Martin Marietta Energy Systems, Inc., for the U.S. Department of Energy under Contract No. DE–AC05–84OR21400, Oak Ridge, Tennessee, 1991.

366. Davenport, J.; Aldykiewicz Jr. A. J.; Isaacs, H. S.; Kendig, M. W.; Mundy, A. M.; *Proc. Symp. On Xray Methods in Corrosion and Interfacial Electrochemistry*, Davenport, A. J.; Gordon II, J. G. (Eds.); Pennington, NJ, The Electrochem Soc., 1992, 92, 1, 306–314.

367. Lake, D. L.; *Industrial Corrosion*, 1989, 7, 4, 12.

368. Opresko, D. M.; "Chemical Hazard Evaluation Group, Biomedical and Environmental Information Analysis Section, Health and Safety Research Division," by Martin Marietta Energy Systems, Inc., for the U.S. Department of Energy under Contract No. DE–AC05–84OR21400, Oak Ridge, Tennessee, 1992.

369. Colturi, T. F.; Kozelski, K. J.; *Mater. Perform.*, 1984, 23, 8, 43.

370. Jones, C. A.; *J. Cooling Tower Inst.*, 1985, 6, 1, 9.

371. Franco, R. J.; Dinielli, N.; Nowicki, R. J.; *Corrosion*, Houston, National Association of Corrosion Engineers, 1985, 85, 133.

372. Weber, T. R.; Stranick, M. A.; Vukasovich, M. S.; *Corrosion, Houston*, National Association of Corrosion Engineers, 1985, 85, 122.

373. Osborne, J. H.; Observations on Chromate Conversion Coatings from a Sol–Gel Perspective, *Progress in Organic Coatings*, 2001, 41, 280–286.

374. Brinker, C. J.; Scherer, G. W.; *Sol–Gel Science*, New York, Academic Press, 1989.

375. Plueddemann, E.; *Silane Coupling Agents*, New York, Plenum Press, 1982.

376. Brinker, C.; Clark, D.; Ulrich, D.; *Better Ceramics through Chemistry Series*, Materials Research Society, 1988.

377. Van Ooij, W. J.; *Symposium: A Systems Approach to Service Life Prediction of Organic Coatings*, Breckrenridge, Colorado, 1997.

378. Van Ooij, W. J.; Song, J.; Subramanian, V.; *Proceedings International Symposium on Aluminum Surfaces Science and Technology*, Antwerp, Belgium, ATB Metallurgie, 1997, XXXVII, 137.

379. Kasten, L. S.; Grant, J. T.; Grebasch, N.; Voevodin, N.; Arnold, F. E.; Donley, M. S.; An XPS Study of Cerium Dopants in Sol–Gel Coatings for Aluminum 2024–T3, *Surface and Coatings Technol.*, 2001, 140, 11.

380. Voevodin, N. N.; Grebasch, N. T.; Soto, W. S.; Arnold, F. E.; Donley, M. S.; Potentiodynamic Evaluation of Sol–Gel Coatings with Inorganic Inhibitors, *Surface and Coatings Technol.*, 2001, 140, 24.

381. Nylund, A.; Chromium–Free Conversion Coatings for Aluminum Surfaces, *Aluminum Transactions*, 2000, 2, 121.

382. Smith, C. J. E.; Baldwin, K. R.; Garrett, S. A.; Gibson, M. C.; Hewins, M. A. H.; Lane, P. L.; *The Development of Chromate–Free Treatments for the Protection of Aerospace Aluminum Alloys*, ATB Metallurgie, 1997, 37, 266.

383. Twite, R.L.; Bierwagen, G.P.; Review of Alternatives to Chromate for Corrosion Protection of Aluminum Aerospace Alloys, *Prog. Org. Coat.*, 1998, 33, 91.

384. Kasten, L. S.; Grant, J. T.; Grebasch, N.; Voevodin, N.; Arnold, F. E.; Donley, M. S.; XPS Study of Cerium Dopants in Sol–Gel Coatings for Aluminum 2024–T3, *Surface and Coatings Technology*, 2001, 140, 11–15.

385. Wilkes, G.L.; Orter, B.; Huang, H.; *Polymer Prep*, 1985, 26, 300.

386. Schmidt, H.; *J. Non–Crystal. Solids*, 1985, 73, 681.

387. Schmidt, H. K.; Aspects of Chemistry and Chemical Processing of Organically Modified Ceramics, *Mater. Res. Soc. Symp. Proc.*, 1990, 180, 961.

388. Mackenzie, J. D.; Structures and Properties of Ormosils, *J. Sol–Gel Sci. Technol.*, 1994, 2, 81.

389. Mackenzie, J. D.; Structures and Properties of Ormosils, *Journal of Sol–Gel Sci. and Technol.*, 1994, 2, 81–86.

390. Brinker, C. J.; Scherer, G. W.; *Sol Gel Science*, San Diego, Academic Press, 1190.

391. Jackson, C. L.; Bauer, B. J.; Nakatani, S. I.; Barnes, J. D.; *Chem. Mater.*, 1996, 8, 727.

392. Wen, J.; Wilkes, G. L.; Organic/Inorganic Hybrid Network Materials by the Sol–Gel Approach, *Chem. Mater.*, 1996, 8, 1667–1681

393. Wen, J.; Wilkes, G. L.; *J. Sol–Gel Sci. Technol.*, 1995, 5, 115.

394. Schmidt, H.; *Mater. Res. Soc. Symp. Proc.*, 1990, 171, 3.

395. Kasemann, R.; Schmidt, H.; *New J. Chem.*, 1994, 18, 1117.

396. Schmidt, H.; Kasemann, R.; Burkhart, T.; Wagner, G.; Arpac, E.; Geiter, E. in *Hybrid Organic–Inorganic Composites*; Mark, J. E., (Ed.); ACS Series 585, Washington, D.C., American Chemical Society, 1995, 331.

397. Tamami, B.; Betrabet, C.; Wilkes, G. L.; *Polym. Bull.*, 1993, 30, 393.

398. Wang, B.; Wilkes, G. L.; *J. Macromol. Sci., Pure Appl. Chem.*, 1994, A31, 249.

399. Betrabet, C.; Wilkes G. L.; *Polym. Prepr.*, 1992, 33, 2, 286.

400. Wen, J.; Wilkes, G. L.; *J. Inorg. Organomet. Polym.*, 1995, 5, 343.

401. Wen, J.; Wilkes, G. L.; *PMSE Prepr.*, 1995, 73, 429.
402. Lebeau, B.; Guermeur, S. C.; *Mater. Res. Soc. Symp. Proc.*, 1994, 346, 315.
403. Sugama, T.; Du Vall, J.E.; *Thin Solid Films*, 1996, 289, 39.
404. Montemor, M. F.; Simoes, A. M.; Ferreria, M. G. S.; Williams, B.; Edwards, H.; The Corrosion Performance of Organosilane Based Pretreatments for Coatings on Galvanized Steel, *Prog. Org.Coat.*, 2000, 38, 17.
405. Metroke, T. L.; Parkhill, R. L.; Knobbe, E.T.; Passivation of Metal Alloys Using Sol–Gel Derived Materials–A Review, *Prog. Org. Coat.*, 2001, 41, 233.
406. Mackenzie, J. D.; Bescher, E. P.; Structures, Properties and Potential Applications of Ormosils, *J. Sol–Gel. Sci. Technol.*, 1998, 13, 371.
407. Guglielmi, M.; Sol–Gel Coatings on Metals, *J. Sol–Gel Sci. Technol.*, 1997, 8, 443.
408. Ooij, W. J.; *Symposium: A Systems Approach to Service Life Prediction of Organic Coatings*, Breckrenridge, CO, 1997.
409. Nylund, A.; Chromium–Free Conversion Coatings for Aluminum Surfaces, *Alum. Trans.*, 2000, 2, 1, 121–137.
410. Ooij, W. J.; "Corrosion Protection of Aluminum Alloys by Conversion Systems and Organic Coatings," *Chemtech*, 1998.
411. "Cleveland Society for Coatings Technology Technical Committee," *Jour. Coat. Tech.*, 1979, 51, 653, 53–57.
412. Walker, P.; *Jour. Coat. Tech.*, 1980, 52, 670, 49–61.
413. Bierwagen, G. P.; Reflections on Corrosion Control by Organic Coatings, *Progress in Organic Coatings*, 1996, 28, 43–48.
414. Kasten, L. S.; Grant, J. T.; Grebasch, N.; Voevodin, N.; Arnold, F. E.; Donley, M. S.; An XPS Study of Cerium Dopants in Sol–Gel Coatings for Aluminum 2024–T3, *Surface and Coatings Technol.*, 2001, 140, 12–13.
415. Metroke, T. L.; Apblett, A.; Corrosion resistance properties of Ormosil coatings on 2024–T3 aluminum, *Proceedings of the 22nd Heat Treating Society Conference and the 2nd International Surface Engineering Congress*, Indianapolis, IN, 327–331, 2003.
416. Pleuddeman, E. P.; *Adhesion Aspects of Polymeric Coatings*, K. Mittal (Ed.), New York, Plenum Press, 1983, 363–377.
417. Metroke, T. L.; Parkhill, R. I.; Knobbe, E. T.; "Synthesis of Hybrid Organic–Inorganic Sol–Gel Coatings for Corrosion Resistance," *Mater. Res. Soc. Symp. Proc.*, 1999, 576, 293.
418. Schmidt, H. H.; Philipp, G.; *J. Non–Cryst. Solids*, 1984, 63, 283.

419. Schmidt, H.; Scholze, H.; Kaiser, H.; *J. Non–Cryst. Solids*, 1984, 63, 1.
420. Schmidt, H.; *Mater. Res. Soc. Symp. Proc.*, 1984, 32, 327.
421. Mackenzie, J. D.; Bescher, E.P., "Structures, Properties, and Potential Applications of Ormosils," *Journal of Sol–Gel Sci. Technol.*, 1998, 13, 371–377.
422. Avnir, D.; Levy, D.; Reisfeld, R.; *J. Phys. Chem.*, 1984, 88, 5956.
423. Pope, J. A.; Asami, A.; Mackenzie, J. D.; *J. Mater. Res.*, 1989, 4, 1018.
424. Wilkes, G. L.; Otter, B.; Huang, H.; *Polymer Prep.*, 1985, 26, 300.
425. Schmidt, H.; *J. Non–Cryt. Solids*, 1985, 73, 681.
426. Mittal, K. L.; Silanes and Other Coupling Agents, *VSP*, 1992.
427. Stratmann, M.; *Adv. Mater.*, 1990, 2, 191.
428. Vetter, K. J.; Schultze, J. W.; *J. Electroanal. Chem.*, 1974, 53, 67.
429. Schultze, J. W.; Koppitz, K. D.; *Electrochim. Acta*, 1976, 21, 327.
430. Hackermann, N.; Hard, R. M.; *1. Int. Congr. Met. Corros.*, London, Butterworths, 1962.
431. Pearson, R. G.; *Science*, 1966, 151, 172.
432. Horner, L.; *Chem. Ztg.*, 1976, 100, 247.
433. Ulman, H.; *Ultrathin Organic Films*, New York, Academic Press, 1993.
434. Wilkes, G. L.; Orler, B.; Huang, H.; *Polym. Prepr.*, 1985, 26, 300.
435. Dislich, H.; *Angew. Chem.*, 1971, 83, 428.
436. Sakka, S.; Kamiya, K.; *J. Non–Crystl. Solids*, 1980, 42, 403.
437. Sakka, S.; *J. Non–Crystl. Solids*, 1985, 73, 651.
438. Klein, L. C.; *Sol–Gel Technology for Thin Films, Fibers, Preforms, Electronics, and Especially Shapes*, Park Ridge, NJ, Noyes Publications, 1988.
439. Yoldas, B. E.; *J. Non–Crystl. Solids*, 1984, 63, 145.
440. Ulrich, D. R.; *Chemtech*, 1988, 18, 242.
441. Brinker, C. J.; Scherrer, G. W.; *Sol–Gel Science, the Physics and Chemistry of Sol–Gel Processing*, San Diego, Academic Press, 1990.
442. Mackenzie, J. D.; Ulrich, D. R. (Eds.).; *Ultrastructure Processing of Advanced Ceramics*, New York, Wiley—Interscience, 1988.
443. Yoldas, B. E.; *J. Mater. Sci.*, 1986, 21, 1086.
444. Sugama, T.; Taylor, C.; Pyrolysis–induced polymetallosiloxane coatings for aluminum substrates, *J. Mater. Sci.*, 1992, 27, 1723.
445. Koehler, E. L.; *Corrosion under Organic Coatings, Proc. U.R. Evans International Conference on Localized Corrosion*, Houston, National Association of Corrosion Engineers, 1971, 117.
446. Guruviah, S.; *J. of the Oil and Colour Chemists' Assoc.*, 1970, 53, 669.

447. Mayne, J. E. O.; *J. of the Oil and Colour Chemists' Assoc.*, 1949, 32, 481.
448. Thomas, A. M.; Gent, W. L.; *Proc. Phys. Soc.*, 1945, 57, 324.
449. Anderson, A. P.; Wright, K. A.; *Industr. Engng. Chem.*, 1941, 33, 991.
450. Edwards, J. D.; Wray, R. I.; *Industr. Engng. Chem.*, 1936, 28, 549.
451. Maitland, C. C.; Mayne, J. E. O.; *Off. Dig.*, 1962, 34, 972.
452. McSweeney, E. E.; *Off. Dig.*, 1965, 37, 626.
453. Wheat, N.; *Prot. Coat. Eur.*, 1998, 3, 24.
454. Hare, C. H.; *Mod. Paint Coat.*, 1986, 76, 38.
455. Boxall, J., *Polym. Paint Colour J.*, 1989, 179, 127.
456. Bieganska, B.; Zubielewicz, M.; Smieszek, E.; *Prog. Org. Coat.*, 1988, 16, 219.
457. Piens, M.; *Evaluations of Protection by Zinc Primers*, Limelette, Belgium, Coat. Res. Inst., 1990.
458. Boxall, J.; *Polym. Paint Colour J.*, 1991, 181, 443.
459. Zimmerman, K.; *Eur. Cot. J.*, 1991, 1, 14.
460. De Lame, C.; Piens, M.; Reactivite de la Poussiere de zinc avec l'oxygene dissous, *Proc. XXIII Fatipec Congress*, Paris, A29–A36, 1996.
461. Forsgren, A.; *Corrosion Control Through Organic Coatings*, Boca Raton, FL, CRC Press, 2006, 115–121.
462. Antropov, L. I.; *Corros. Sci.*, 1967, 7, 607.
463. Iofa, Z. A.; Batrakov, V. V.; Ba, ChoNgok; *Electrochim. Acta*, 1964, 9, 1645.
464. Scully, J. C.; *Corros. Sci.*, 1968, 8, 513.
465. McBee, C. L.; Kruger, J.; *Proceedings, U. R. Evans, International Conference on Localized Corrosion*, 1971, Houston, National Association of Corrosion Engineers, 1974, 252.
466. Lin, L. F.; Chao, C. Y.; MacDonald, D. D.; *J. Electrochem. Soc.*, 1981, 128, 1194.
467. Abd El Kader, J. M.; El Warraky, A. A.; Abd El Aziz, A. M.; Corrosion Inhibition of Mild Steel by Sodium Tungstate in Neutral Solution, *British Corrosion Journal*, 1998, 33, 2, 152–157.
468. Lutey, R.; *Microbiological Corrosion*, Houston, National Association of Corrosion Engineers, 1980, 80, 39.
469. AISC Specifications, American Institute of Steel Construction, 13th ed., 2005.
470. Waseda, Y.; Suzuki, S.; *Characterization of Corrosion Products on Steel Surfaces*, Berlin, Springer Berlin Heidelberg 2006, 294.
471. ACI Committee 222; *Corrosion of Metals in Concrete (ACI222R96)*, Farmington Hills, MI, American Concrete Institute, 1996, 30.
472. Gjørv, Odd E.; "Durability of Concrete Structures in the Ocean Environment," *Proceedings, FIP Symposium on Concrete Sea*

Structures (Tbilisi, Sept. 1972), London, Federation Internationale de la Precontrainte, 1973, 141–145.

473. Foley, R. T., "Complex Ions and Corrosion," *Journal of Electrochemical Society*, 1975, 122, 11, 1493–1549.

474. Sluijter, W. L., and Kreijger, P. C., "Potentio Dynamic Polarization Curves and Steel Corrosion," *Heron* (Delft), 1977, 22, 1, 13–27.

475. Weise, C. H., "Determination of the Free Calcium Hydroxide Contents of Hydrated Portland Cements and Calcium Silicates," *Analytical Chemistry*, 1961, 33, 7, 877–882.

476. Mehta, P. K., "Effect of Cement Composition on Corrosion of Reinforcing Steel in Concrete," *Chloride Corrosion of Steel in Concrete*, STP–629, Philadelphia, ASTM, 1977, 12–1.

477. ACI Committee 201, *Guide to Durable Concrete* (ACI 201.2R77) *(Reaffirmed 1982)*, Detroit, American Concrete Institute, 1977, 37.

478. Beaton, J. L., and Stratfull, R. F., "Environmental Influence on Corrosion of Reinforcing in Concrete Bridge Substructures," *Highway Research Record* No. 14, Highway Research Board, 1963, 60–78.

479. Ost, Borje, and Monfore, G. E., "Penetration of Chloride into Concrete," *Journal, PCA Research and Development Laboratories*, 1966, 8, 1, 46–52.

480. Clear, K. C., "Time-to-Corrosion of Reinforcing Steel in Concrete Slabs, V. 3: Performance after 830 Daily Salt Applications," *Report No. FHWA–RD–76–70*, Washington, D.C., Federal Highway Administration, 1976, 64.

481. *Building Code Requirements for Structural Concrete and Commentary (ACI 318M–08)*, Farmington Hills, MI, American Concrete Institute, 2008.

482. BS8110–85 *The Structural Use of Concrete*, London, British Standards Institution.

483. Boulware, R. L., and Elliott, A. L., "California Seals Salt-Damaged Bridge Decks," *Civil Engineering–ASCE*, 1971, 41, 10, 42–44.

484. Van Til, C. J.; Carr, B. J.; and Vallerga, B. A., "Waterproof Membranes for Protection of Concrete Bridge Decks: Laboratory Phase," NCHRP Report No. 165, Washington, D.C., Transportation Research Board, 1976, 70.

485. Frascoia, R. I., "Vermont's Experience with Bridge Deck ProtectiveSystems," *Chloride Corrosion of Steel in Concrete*, STP–629, Philadelphia, ASTM, 1977, 69–81.

486. Smoak, W. G., "Development and Field Evaluation of a Technique for Polymer Impregnation of New Concrete Bridge Deck Surfaces," Report No. FHWARD795, Washington, D.C., Federal Highway Administration, 1976.

487. Tremper, Bailey, "Repair of Damaged Concrete with Epoxy Resins,"ACI JOURNAL, *Proceedings,* 1960, 57, 2, 173–182.

488. McConnell, W. R., "Epoxy Surface Treatments for Portland Cement Concrete Pavements," *Epoxies with Concrete,* SP–21, Detroit, American Concrete Institute, 1968, 9–17.

489. McKeel, W. T., Jr., and Tyson, S. S., "Two–Course Bonded Construction and Overlays," *ACI Journal, Proceedings,* 1975, 72, 12, 708–713.

490. Verbeck, George J., "Mechanisms of Corrosion of Steel in Concrete," *Corrosion of Metals in Concrete,* SP49, Detroit, American Concrete Institute, 1975, 21–38.

491. Cook, A. R., and Radtke, S. F., "Recent Research on Galvanized Steel for Reinforcement of Concrete," *Chloride Corrosion of Steel in Concrete,* STP629, Philadelphia, ASTM, 1977, 51–60.

492. Arnold, C. J., "Galvanized Steel Reinforced Concrete Bridge Decks: Progress Report," *Report No. FHWA–MI–78–R1033,* Washington, D.C., Federal Highway Administration, 1976.

493. Clear, K. C., "Time–to–Corrosion of Reinforcing Steel in Concrete Slabs, V. 4: Galvanized Reinforcing Steel," *Report No. FHWA–RD–82–028,* Washington, D.C., Federal Highway Administration, 1981.

494. Johnston, D. W., and Zia, P., "Bond Characteristics of Epoxy Coated Reinforcing Bars," *Report No. FHWA–NC–82–002,* Washington, D.C., Federal Highway Administration, 1982.

495. I. A. Callander and F. Gianettl, "A review on the Use of Calcium Nitrite Corrosion Inhibitor to Improve the Durability of Reinforced Concrete", *V. Corrosion Symposium,* 1996, 139–149.

496. "Corrosion and Repair of Unbonded Single Strand Tendons," *ACI/ASCE Committee Report 423, 4R–98,* American Concrete Institute, 1998, 20.

497. Stratfull, R. F., "Experimental Cathodic Protection of a Bridge Deck," *Transportation Research Record No. 500,* Transportation Research Board, 1974, 1–15.

498. Fromm, H. J., "Cathodic Protection of Rebar in Concrete Bridge Decks," *Materials Performance,* 1977, 16, 11, 21–29.

499. Robinson, R. C., "Cathodic Protection of Steel in Concrete," *Corrosion of Metals in Concrete,* SP49, Detroit, American Concrete Institute, 1975, 83–93.

500. Stratfull, Richard F., "Criteria for the Cathodic Protection of Bridge Decks," *Corrosion of Reinforcement in Concrete Construction*, Chichester, Ellis Horwood, 1983, 287–331.

501. Uhlig, Herbert H., *Corrosion and Corrosion Control*, New York, John Wiley & Sons, 1963.

502. Fontana, M. G.; Greene, N. D.; *Corrosion Engineering*, 2nd Ed., New York, McGraw–Hill 1978, 448.

503. Chaudhary, Z., "Design and Protection Criteria for Cathodic Protection of Seawater Intake Structures in Petrochemical Plants," *Marine Corrosion in Tropical Environments, ASTM STP 1399*, Dean, S.W.; Hernandez–Duque Delgadillo, G.; Bushman, J.B. (Eds.); , West Conshohocken, PA, American Society for Testing and Materials, 2000.

504. Dreyman, E. W., "Cathodic Protection of Structures in Coral Sands in the Presence of Salt Water," *Marine Corrosion in Tropical Environments*, ASTM STP 1399, Dean, S.W.; Hernandez–Duque Delgadillo, G.; Bushman, J.B. (Eds.); West Conshohocken, PA, American Society for Testing and Materials, 2000.

505. Tuttle, R. N.; "Corrosion in Oil and Gas Production," *Journal of Petroleum Technology*, July 1987, 756–762.

506. Hashen, A.; Riad, W.;"The Influence and Industrial Atmospheres on the Corrosion of Carbon Steel, Galvanized Steel and Copper in Kuwait," *V. Corrosion Symposium*, Nov. 1996, 53–62

507. Masaden, S.; "Corrosion and Rehabilitation of Steel Reinforced Concrete Structure Exposed to Dead Sea", *International Corrosion symposium*, Nov. 2006, 198–206.

Index

Also of Interest

Check out these other related titles coming soon from Scrivener Publishing

An Introduction to Petroleum Technology, Economics, and Politics, by James Speight, September 2011, ISBN 9781118012994. The perfect primer for anyone wishing to learn about the petroleum industry, for the layperson or the engineer.

Ethics in Engineering, by James Speight and Russell Foote, ISBN 9780470626023. Covers the most thought-provoking ethical questions in engineering. *NOW AVAILABLE!*

Formulas and Calculations for Drilling Engineers, by Robello Samuel, ISBN 9780470625996. The most comprehensive coverage of solutions for daily drilling problems ever published. *NOW AVAILABLE!*

Emergency Response Management for Offshore Oil Spills, by Nicholas P. Cheremisinoff, PhD, and Anton Davletshin, ISBN 9780470927120. The first book to examine the Deepwater Horizon disaster and offer processes for safety and environmental protection. *NOW AVAILABLE!*

Advanced Petroleum Reservoir Simulation, by M.R. Islam, S.H. Mousavizadegan, Shabbir Mustafiz, and Jamal H. Abou-Kassem, ISBN 9780470625811. The state of the art in petroleum reservoir simulation. *NOW AVAILABLE!*

Energy Storage: A New Approach, by Ralph Zito, ISBN 9780470625910. Exploring the potential of reversible concentrations cells, the author of this groundbreaking volume reveals new technologies to solve the global crisis of energy storage. *NOW AVAILABLE!*

Zero-Waste Engineering, by Rafiqul Islam, February 2012, ISBN 9780470626047. In this controvercial new volume, the author explores the question of zero-waste engineering and how it can be done, efficiently and profitably.

Fundamentals of LNG Plant Design, by Saeid Mokhatab, David Messersmith, Walter Sonne, and Kamal Shah, August 2012. The only book of its kind, detailing LNG plant design, as the world turns more and more to LNG for its energy needs.

Flow Assurance, by Boyun Guo and Rafiqul Islam, January 2012, ISBN 9780470626085. Comprehensive and state-of-the-art guide to flow assurance in the petroleum industry.

Advances in Natural Gas Engineering Series (3 volumes):

Acid Gas Injection and Related Technologies, by Ying Wu and John J. Carroll, ISBN 9781118016640. The only book covering this timely topic in natural gas. *NOW AVAILABLE!*

Carbon Dioxide Sequestration and Related Technologies, by Ying Wu and John J. Carroll, ISBN 9780470938768. volume two focuses on one of the hottest topics in any field of engineering, carbon dioxide sequestration. *NOW AVAILABLE!*

Sour Gas and Related Technologies, by Ying Wu and John J. Carroll, ISBN 9780470948149. This third volume in the series focuses on sour gas, one of the most important issues facing chemical and process engineers. *PUBLISHING MAY 2012.*